PRESSURE DIECASTING

Part 1

Metals — Machines — Furnaces

OTHER TITLES IN THE SERIES

NOTICE TO READERS

PRESSURE DIECASTING

Part 1

Metals — Machines — Furnaces

B. UPTON, M.Tech., C. Eng., M.I.M.
Buhler-Miag (England) Ltd

PERGAMON PRESS

OXFORD · NEW YORK · TORONTO · SYDNEY · PARIS · FRANKFURT

U.K.   Pergamon Press Ltd., Headington Hill Hall,
                        Oxford OX3 0BW, England

U.S.A.   Pergamon Press Inc., Maxwell House, Fairview Park,
                        Elmsford, New York 10523, U.S.A.

CANADA   Pergamon Press Canada Ltd., Suite 104,
                        150 Consumers Road, Willowdale, Ontario M2J 1P9, Canada

AUSTRALIA   Pergamon Press (Aust.) Pty. Ltd., P.O. Box 544,
                        Potts Point, N.S.W. 2011, Australia

FRANCE   Pergamon Press SARL, 24 rue des Ecoles,
                        75240 Paris, Cedex 05, France

FEDERAL REPUBLIC   Pergamon Press GmbH, 6242 Kronberg-Taunus,
OF GERMANY   Hammerweg 6, Federal Republic of Germany

Copyright © 1982 Pergamon Press Ltd.

First edition 1982

Library of Congress Cataloging in Publication Data
Upton, B.
Pressure diecasting.
(Materials engineering practice)
Contents: pt. 1. Metals, machines, furnaces.
1. Die-casting. I. Title. II. Series.
TS239.U67   1982   671.2'53   81-15707
AACR2

British Library Cataloguing in Publication Data
Upton, B.
Pressure diecasting.—(Materials engineering practice)
Part 1: Metals, machines, furnaces
1. Die-casting
I. Title II. Series
671.2'53   TS239
ISBN 0-08-027621-0 (Hardcover)
ISBN 0-08-027622-9 (Flexicover)

Printed in Great Britain by A. Wheaton & Co. Ltd., Exeter

Materials Engineering Practice

FOREWORD

The title of this new series of books "Materials Engineering Practice" is well chosen since it brings to our attention that in an era where science, technology and engineering condition our material standards of living, the effectiveness of practical skills in translating concepts and designs from the imagination or drawing board to commercial reality is the ultimate test by which an industrial economy succeeds.

The economic wealth of this country is based principally upon the transformation and manipulation of *materials* through *engineering practice.* Every material, metals and their alloys and the vast range of ceramics and polymers has characteristics which requires specialist knowledge to get the best out of them in practice, and this series is intended to offer a distillation of the best practices based on increasing understanding of the subtleties of material properties and behaviour and on improving experience internationally. Thus the series covers or will cover such diverse areas of practical interest as surface treatments, joining methods, process practices, inspection techniques and many other features concerned with materials engineering.

It is to be hoped that the reader will use this book as the base on which to develop his own excellence and perhaps his own practices as a result of his experience and that these personal developments will find their way into later editions for future readers. In past years it may well have been true that if a man made a better mousetrap the world would beat a path to his door. Today however to make a better mousetrap requires more direct communication between those who know how to make the better mousetrap and those who wish to

v

know. Hopefully this series will make its contribution towards improving these exchanges.

MONTY FINNISTON

Contents

Chapter 1

Introduction to Pressure Diecasting

Technicians are by definition concerned with the technical aspects of the industry in which they are working, and this infers a study of the industry at the present time together with keeping an eye on future developments. The historical development of a process, its place in the economy and its competing processes are frequently of little concern to the technician. However, the survival of any manufacturing process depends upon the latter two factors as much as upon its own technical excellence, so no apology is offered for commencing a technical book with a chapter having no direct relevance to the work a technician is expected to do.

1. BRIEF HISTORY

Pressure diecasting belongs to the family of casting processes which utilise a permanent mould. Other processes in this family include gravity diecasting and low-pressure diecasting. Historically, gravity diecasting predates the other two processes by several thousand years. Archaelogical evidence would suggest that a form of gravity diecasting was in use during the Bronze Age period for the manufacture of axeheads, the mould material being stone. Our ancestors realised that a permanent mould process offered an advantage where large numbers of castings to the same design had to be made, and the same considerations apply in the modern world.

Like many technological processes, the origins of pressure diecasting in more modern times are not known with any degree of certainty, but it is generally considered that they occurred in the early part of the nineteenth century in connection with the production of printer's type, using the lead-tin alloy common in that industry.

As early as 1822 the process was showing its production potential.[1]* Dr. William Church had produced a typecasting machine with an output of 12,000-20,000 letters per day and an automatic machine produced by David Bruce Jr. in 1838 had an output of 165 thin letters per minute.

It would not be surprising if these early machines differed considerably in appearance from their modern derivatives. Sturgiss[2] in 1849 patented a machine (Fig 1.1) in which molten metal in the lower chamber was forced through an inclined port and out of the nozzle into the die by the central ram actuated by a lever. The plug and sprue cut-off was then advanced into the nozzle, at the same time opening a port which admitted molten metal from the upper chamber to the inclined port to refill the lower chamber as the plunger was elevated. A similar system was used by Dusenbery[1] in 1877, but this time a hollow plunger with a valve to admit molten metal from an upper chamber to a lower chamber was involved (Fig. 1.2). The design of the first gooseneck machine is credited to Van Wagner in 1907[2] (Fig. 1.3). This machine was designed for pneumatic operation

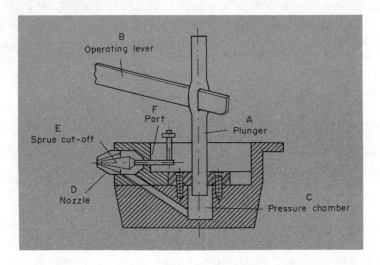

Fig. 1.1 Diagram of a Portion of the Sturgiss Machine Showing, Partially in Section, the Pressure Chamber, Plunger and Nozzle (1849).

and had no plunger. The gooseneck was fixed and had to be filled by hand ladling of the metal after the die had been removed. Any history of the diecasting process must mention the work of Charles Babbage.[1] Babbage was the designer of the first mechanical computer in the 1860s and being faced with the need for a larger number

*Superscript numbers refer to References at end of chapter.

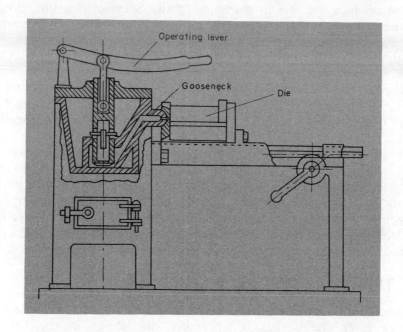

FIG. 1.2   DUSENBERY'S DIECASTING MACHINE OF 1877, WITH HAND LEVER INJECTION.

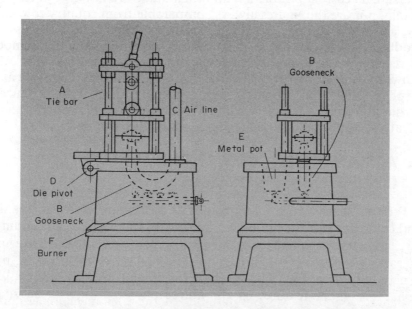

FIG. 1.3   DIAGRAM OF VAN WAGNER GOOSENECK MACHINE (1907).

of repetitive castings turned to the diecasting process. Several of these castings are still in existence in the Science Museum, South Kensington, London. It was reported at the time that the alloy used for these castings was pewter hardened with zinc, but spectrographic analysis of three samples has shown this not to be the case. One casting was found to be a lead base alloy with a nearly eutectic antimony content and only 1% zinc, whilst two others were found to be tin-antimony alloys of the Britannia metal type. Britannia metals were common in Victorian England for the production of decorative wear.

The latter part of the nineteenth century saw increasing use of diecastings in the manufacture of cash registers, phonographs and bicycles, but the year 1904 is possibly the most significant date in the history of diecasting since it is then that the H. H. Franklin Company began to manufacture diecast bearings for automobile connecting rods,[2] thus beginning the association of diecastings with the automobile industry which has gone on since that date.

2. TYPICAL USES

The dominance of the automobile industry is shown in Table 1.1, which gives the outlets for pressure diecastings in the United Kingdom in the year 1971. The types of castings made are shown in Table 1.2 for the same year. International statistics related to pressure diecasting production and their value are not always easy to obtain, nor are the figures always comparable from country to country, and hence can be misleading. An idea of the pressure diecasting production potential of a country, can be obtained from the number of machines in use and an estimate produced in 1971 (Table 1.3) shows the dominance of the United States. This should not come as a surprise, since that country is the home of the mass-produced automobile and of a society committed to the purchase of consumer durables, both markets being vast users of diecastings.

3. ADVANTAGES, COSTS AND COMPETITIVE PROCESSES

Pressure diecasting is the fastest of all casting processes and ranks and competes with stamping, die forging, plastic injection moulding, etc., where large numbers of identical parts are required. Pressure diecasting is also a competitor with sand and gravity diecasting, but generally each process has certain inherent advantages and limitations. The main features of the above processes have been summarised[3] (Tables 1.4 and 1.5) and the following conclusions have been drawn with regard to pressure diecasting and its applications.

TABLE 1.1. THE OUTLETS FOR PRESSURE DIECASTING IN THE U.K. 1971

Industry	Aluminium(%)[a]	Zinc(%)[b]
Automotive	55	30
Electrical engineering	13	NA
Mechanical engineering	9	NA
Domestic and office equipment	14	30
Toys	—	15
Others	9	25

NA, not available.

Sources: [a]Aluminium Federation.
[b]Estimates provided by the Zinc Development Association.

TABLE 1.2. TYPICAL APPLICATIONS FOR PRESSURE DIECASTING IN THE U.K. 1971.

Aluminium	Zinc
Automotive	
Radiator grills	Door handles
Carburettor bodies	Carburettor bodies
Distributor bodies	Lamp bezels
Water pump housings	Distributor bodies
Engine blocks	Dynamo end plates
Cylinder heads	Starter motor end plates
Housings for clutch,	Thermostat housings
gearbox, timing gears, etc.	Wiper motor casing
Electrical engineering	
Motor frames	Switchgear housings
Motor rotors	and covers
Mechanical engineering	
Diesel engine oil pans	Gearbox housings and
	covers
Pulleys and other rotating	Pull handles
parts where weight is	
critical	
	Pulleys
Domestic and office equipment	
Portable tool covers and	Washing machine pumps
frames	Refrigerator door handles
Typewriter frames	Record player decks and arms
"Iron" faceplates	TV bezels (chromium
TV bezels	plated)
Washing machine paddles	

TABLE 1.3 DIECASTING MACHINE POPULATION UP TO SEPTEMBER 1971, WESTERN WORLD

Country	Total No. of machines Estimated	Estimated Average tonnage of Material used per machine per year
U.S.A.	15,000	85
Canada	650	88
Germany	2600	88
U.K.	1750	85
Italy	2100	62
France	1250	71
Japan	3900	73

TABLE 1.5 COST COMPARISON

Process	Appearance and finish	Cost
Pressure diecasting	Very good, can be finished with variety of mechanical, plated, chemical or organic finishes	High equipment cost, high tool cost, low piece cost on high-activity items
Gravity diecasting	Usually machined or ground but left with base metal surface	Low equipment cost, medium die cost. Piece cost between diecasting and sand casting
Sand casting	Inferior to die or gravity castings, machined or ground, left with base metal finish	Low equipment cost, low tool cost, high piece cost
Powder metal	Good, but porous, usually left with base metal finish	Medium equipment and tool cost, low unit cost on high-activity items
Drop forging and die pressing	Inferior to diecasting, usually left in base metal finish	High equipment cost, high tool cost, reasonably low unit cost on high-activity items
Plastic moulding	Excellent, can be moulded in variety of colours with excellent finish	High equipment cost, high tooling cost, low piece cost, on high-activity items

TABLE 1.4 COMPARISON OF DIECASTING WITH OTHER PRODUCTION PROCESSES

Process	Materials	Rate of production	Strength of parts	Complexity
Pressure diecasting	Lead, tin, zinc, magnesium, aluminium, copper alloys	Very high: on certain components up to 2000 per hour	High unit strength	From simple to very complex
Gravity diecasting	Iron, magnesium, aluminium, copper alloys	Relatively low	High, particularly when heat-treated	In general not so complex as diecastings unless sand cores are used
Sand casting	Principally iron, magnesium, aluminium and copper alloys	Low — unless foundries fully mechanised	Less than that of diecastings and gravity diecasting	From simple to very complex
Drop forging and die pressing	Steel, magnesium, aluminium and copper alloys	Reasonable	Highest of all	Fairly complex
Plastic moulding	Thermoplastic and Thermosetting resins	Very high — equal to diecastings	Less than parts mentioned above	As diecastings with the advantages that components are self-coloured

A pressure diecasting may suit the application best if the answer to any one of the following questions is yes:

1. Is the quantity involved considerable (5000 or more)? Pressure diecasting is characteristically a high-speed production process. For this reason it is naturally suited to economical production on a quantity basis.
2. Is the tooling for machining expensive? Pressure diecastings usually require very few secondary operations. Reduction of tooling costs means an important saving in overall production expense. An order for a relatively small number of castings frequently justifies pressure diecasting, because the need for costly machine tools is eliminated.
3. Does machining, assembling or surface finishing account for an appreciable part of the cost of the finished article? The use of pressure diecastings reduces the time and expense required to complete the product. Assembly is rapid because light metal pressure diecastings have uniform and repeatable dimentions. Surfaces are smooth as cast, so that many finishing operations are simplified or eliminted.
4. Is a reduction of the investment in machine tools and plant floor space desirable? Frequently the use of pressure diecastings is very advantageous as a measure to conserve plant facilities. Castings are supplied with the flash trimmed off, and in many cases they are almost ready for assembly. Elimination of the multitude of operations that would be required with other methods can free productive capacity and space for other uses.

It was also[3] pointed out that the ability to produce undercuts and intricate holes by means of sliding die members was also a considerable advantage of the diecasting process.

Typical properties of diecastings and competitive production routes and materials are given in Table 1.6[4]. Table 1.6 is not an exhaustive list of factors or advantages and obviously for any component the competing routes must be studied, costs worked out and intangible factors evaluated. Such exercises have been carried out in the United Kingdom[5] and in South Africa[6]

Cost figures for gravity and pressure routes are quoted[5] for the production of large, medium and smaller castings and it was suggested that there was likely to be a break-even point in casting size between these two processes. High-pressure diecastings would have a price advantage for castings made on machines up to 800-1200 tons locking force, providing the quantities were correct.

TABLE 1.6 COMPARATIVE CHART OF MATERIAL PROPERTIES

Material description	Tensile strength N/mm² room temp.	Tensile modulus N/mm² room temp.	Tensile elongation % room temp.	Impact resistance J	Hardness	Specific gravity
ABS (acrylonitrile butadiene styrene)	37-59	1400-2700	5-40	0.8-2	80-120R	1.02-1.2
Acetal (homopolymer)	68	2800-3400	15-75	0.9-1.6	120R	1.42
Polyamide nylon 6/6	61-80	1200-2700	60.300	0.6-1.4	108-118R	1.14
Polycarbonate	54-64	2400	60-100	8	118R	1.2
Polypropylene (homopolymer)	29-37	1100-1500	200-700	0.4-1	80-110R	0.9
Polystyrene (high impact)	17-33	1400-3000	1.2-4.5	0.6-5.5	110-120R	1.05
Zinc alloy BS1004 alloy A (die cast)	247-286	100,000	15-25	58	65-83B	6.7
Zinc alloy BS1004 alloy B (die cast)	292-335	100,000	9-14	57	74-72B	6.7
Aluminium alloy BS1490 LM24 (die cast)	320	71,000	1-3	3.4	85B	2.8
Magnesium alloy BS2970 MAG7 (die cast)	220-248	44,000	2-5	27	55-67B	1.8
Brass BS1400 PCB1 (die cast)	280-370	105,000	25-40	33	60-70B	8.3
Steel BS970 (hot forged)	850	206,000	18-20	41 (minimum)	280B	7.8
Brass BS2872 CZ122 (hot forged)	400	105,000	25-35	34	80-100B	8.4

TABLE I. 6 (Cont.)

Material description	Electrical properties	Thermal conductivity W/m°C	Linear expansion coefficient per°C	Specific heat	Heat deflection temperature[b] °C	Flammability
ABS (acrylonitrile butadiene styrene)	10^{24} ohm/m/mm²	1.9×10^{-1} -3.4×10⁻¹	5.5×10^{-5} -11×10^{-5}	0.3-0.4	80-90	Slow. No drip
Acetal (homopolymer)	5.3×10^{21} ohm/m/mm²	2.3×10^{-1}	8.1×10^{-5}	0.35	110 (copolymer)	Slow. 20mm/min
Polyamide nylon 6/6	10^{23} ohm/m/mm²	2.45×10^{-1}	8.0×10^{-5}	0.4	70-80	Self-extin-guishing
Polycarbonate	10^{24} ohm/m/mm²	1.9×10^{-1}	6.6×10^{-5}	0.3	130	Self-extin-guishing
Polypropylene (homopolymer)	10^{24} ohm/m/mm²	1.2×10^{-1}	5.8×10^{-5} -10.2×10^{-5}	0.46	60	Slow
Polystyrene (high impact)	10^{24} ohm/m/mm²	4.2×10^{-1} -12.6×10^{-1}	3.4×10^{-5} -7×10^{-5}	0.32-0.37	80	Slow
Zinc alloy BS1004 alloy A (die cast)	15.7 ohm⁻¹/m/ mm²	1.13×10^{2}	2.7×10^{-5}	0.093	205	Nil
Zinc alloy BS1004 alloy B (die cast)	15.3 ohm¹⁻/m/ mm²	1.09×10^{2}	2.7×10^{-5}	0.093	—	Nil

TABLE I. 6 *(Cont.)*

Material description	Electrical properties	Thermal conductivity W/m/°C	Linear expansion coefficient per °C	Specific heat	Heat deflection temperature[b] °C	Flammability
Aluminium alloy BS1490 LM24 (die cast)	14 ohm^{-1}/m mm^2	2.4×10^2	2.3×10^{-5}	0.21	a	Nil
Magnesium alloy BS2970 MAG7 (die cast)	18 ohm^{-1}/m mm^2	1.5×10^2	2.7×10^{-5}	0.245	a	Rapid in finely divided form
Brass BS1400 PCB1 (die cast)	11 ohm^{-1}/m mm^2	0.8×10^2	2.1×10^{-5}	0.092	a	Nil
Steel BS970 (hot forged)	8.7 ohm^{-1}/m mm^2	0.5×10^2	1.24×10^{-5}	0.105	a	Nil
Brass BS2872 CZ122 (hot forged)	15.1 ohm^{-1}/m mm^2	1.05×10^2	2.0×10^{-5}	0.092	a	Nil

[a] These alloys can be used under moderate stress at temperatures up to about 200°C, depending on their heat treatment, etc. Advice should be sought on high temperature applications of any metals.

[b] BS2782 Method 102G. Stress 1.81N/mm^2. This test method is not normally used for metals and data for zinc alloys are included for comparison only.

TABLE I. 6 *(Cont.)*

Material description	Water absorption % at room temp in 24 hr	Chemical resistance				Organic solvents	Price New pence/ kg (Mid-1972)
		Acids		Alkalis			
		Weak	Strong	Weak	Strong		
ABS (acrylonitrile butadiene styrene)	0.2-0.4 (<1.0/year)	G	G	G	G	P	34
Acetal (homopolymer)	not available	F	P	F	P	G	66
Polyamide nylon 6/6	1.5 (hygroscopic)	G	P	G	F	G	60
Polycarbonate	0.15	G	G	G	F	P	68
Polypropylene (homopolymer)	0.03	G	G	G	G	G	22
Polystyrene (high impact)	0.05-0.6	G	G	G	G	P	17
Zinc alloy BS1004 alloy A (die cast)	Nil	P	P	G	P	G (anhydrous solvents)	18
Zinc alloy BS1004 alloy B (die cast)	Nil	P	P	G	P	G (anhydrous solvents)	19
Aluminium alloy BS1490 LM24 (die cast)	Nil	P	P	F	P	G	18
Magnesium alloy BS2970 MAG7 (die cast)	Nil	P	P	F	G	G or F	31
Brass BS1400 PCB1 (die cast)	Nil	G or F	G or F	G (not ammonia)	G	G	35
Steel BS970 (hot forged)	Nil	P	P	F	G	G (anhydrous solvents)	12
Brass BS2872 CZ122 (hot forged)	Nil	G or F	G or F	G (not ammonia)	G	G	44

G-good f-fair p-poor

Further cost analysis is given in Table 1.7[6] where the detail required for accurate cost assessment will readily be seen. These figures have been used to compare the production costs of a connecting rod and housing from the automobile industry made by gravity diecasting, drop forging and pressure diecasting (Table 1.8). Such exercises can be useful, but the cost information can become quickly out of date. Detailed study must also be given to the data to establish whether the conditions quoted are the same as one's own. This is particularly important in the field of labour costs. Labour costs of £1 sterling per hour for two non-European workers[6] are unlikely to be duplicated in Western Europe.

TABLE 1.7 PRESSURE DIECASTING PRODUCTION COSTS

			1R = 62.5p	Remarks
A^1	Investment of the diecasting machine	R	60,000	complete, ready for connection and for production
A^2	Import costs	R	7000	fob, freight, packing, clearing costs
A^3	Die costs	R	6800	housing + conecting rod
A^4	Furnace	R	5000	
B	Installation costs	R	1000	
C	Costs for foundation	R	400	
D	Depreciation of investment	a	5	
E	Rate of interest	%/a	12	
F	Maintenance of machine and accessories	%/a	5	
G^1	Local requirement	m^2	25	
G^2	Costs of space per machine unit	$R/m^2.a$	15	
H	Power consumption per hour	kW/h	16	40% of the installed capacity
I	Costs for power consumption	R/kWh	0.02	
J	Production time per annum	h/a	1450	1 shift, 65.9% rate of utilization
K	Labour costs	R/h	1.60	2 non-European operators
L^1	Overheads	%	400	
M	Total investment, inclusive of installation and foundation	R	80,200	$A+B+C$
N	Amount of depreciation	R/a	16,040	M:D
O	Interest	R/a	4812	M.E:100.0,5
P^1	Repair and maintenance costs	R/a	3940	A.F 100
P^2	Working expenses	R/a	2500	water, air, oil, nitrogen, die release agents
Q	Space costs	R/a	375	$G^1.G^2$
R^1	Machine costs, per annum	R/a	27,667	$N+O+P^1+P^2+Q$
R^2	Machine costs, per hour	R/h	19.08	R^1:J
S	Energy costs	R/h	0.32	H.I
L^2	Overheads	R/h	6.40	$K.L^1$:100
T	Costs for one production hour	R/h	27.40	$R^2+S+K+L^2$
U	Useful casting per hour housing connecting rod		163 86	
V	Mass of casting (kg) housing connecting rod		0.46 0.42	
W	Material costs (Rand) housing connecting rod		0.51 0.466	LM6:R 1,11/Kg
X	Production costs of one casting (R) housing connecting rod		0.68 0.79	T : U + W

TABLE 1.8 COST COMPARISON OF VARIOUS PRODUCTION ROUTES

| | Housing | | Dropforging | Connecting rod |
	Gravity	Pressure diecasting		Pressure diecasting
Weight of casting (kg)	0.55	0.46	0.42	0.42
saving (%)		16.4		-no-
Metal price LM24 (R/kg)				
LM6 (R/kg)		1.11		
1474 HE 20 (R/kg)			1.95	
Material costs per casting (R)	0.51	0.51	0.82	0.47
saving (%)		-no-		60.2
Processing cost (R)	0.22	0.17	0.11	0.32
saving (%)		22.7		-no-
Cleaning and machining (R)	0.09	0.03	0.02	0.02
saving (%)		6.66		-no-
Total Costs (R)	0.82	0.71	0.95	0.81
Saving per casting (R)		0.11		0.14
Saving per casting (%)		13.4		14.7
Savings over One Year (R)		10,340.00		10,080.00

4. FUTURE

There is little doubt that the use of pressure diecastings in any country is related to living standards, and since standards tend to rise, the overall future of pressure diecasting would appear to be assured. What is not certain is the relative position of the metals being cast. Figure 1.4 shows production figures for zinc and aluminium pressure diecasting in the United Kingdom and the United States and whilst a steady growth is seen for aluminium, zinc has been relatively static since the middle 1960s in both countries. The reason for this can be found by studying the competitive processes. Zinc castings appear more vulnerable to substitution than do those made in aluminium. Irrespective of the alloy being cast, the future of the process is bound to depend upon the ability of the industry to mechanise, to automate, to obtain greater productivity from existing plant and above all to achieve success in technology transfer. In all these fields the role of the technician can be vital.

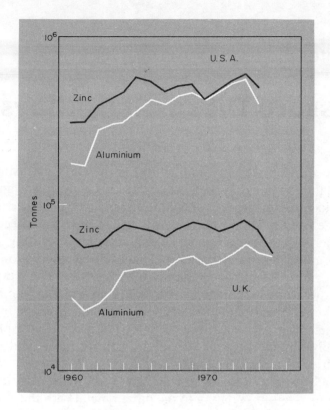

FIG. 1.4 DIECASTINGS PRODUCTION

REFERENCES

1. H. K. Barton, Charles Babbage and the beginnings of diecasting, *Machinery and Production Engineering,* 27 Oct. 1971.

2. *Diecasting for Engineers.* New Jersey Zinc Company, New York, U.S.A.

3. A. G. Wakefield, The role of diecasting in industry, *Proceedings of the 1st International Conference on Production Development and Manufacturing Technology,* University of Strathclyde, Sept. 1969, pp.231-241.

4. *Zinc Diecasting Guide.* Zinc Development Association, 34 Berkeley Square, London.

5. S. D. Apsley Gravity versus high pressure routes to the economical production of large castings. Diecasting Society's National Conference, May 1977.

6. S. J. Chilla, The economy of pressure diecasting, Founding, Welding, Production Engineering Journal, March 1977. pp.33-42.

Chapter 2

Pressure Diecasting Alloys

It has been mentioned in Chapter 1 that the early diecastings were made in the low melting-point alloys based on lead, tin, antimony, etc., which are not aggressive to the materials of construction from which the early machines were made. These have now been largely superseded, as the requirements for "structural" casting have increased, and for most practical purposes diecasting is confined to alloys of zinc, aluminium, magnesium and copper. There are, nevertheless, diecastings still being made in the low melting-point alloys and at the other extreme, diecasting are now being made in steel.

The metallurgy of the various diecasting alloys and their properties are readily obtained from textbooks on the subject, and no more than a brief mention is necessary here.

The alloys are adequately described by the simple binary equilibrium diagrams relating to the parent metal with the chief alloying element, and since the opportunities for heat treatment and joining by other than adhesives or mechanical methods is limited, the alloys used pose few metallurgical problems. The U.K. specifications are given in Table 2.1 and the relevant portions of the thermal equilibrium diagrams for the alloys of interest are given in Figs. 2.1 -2.4. It should be remembered that the cooling rates achieved in diecasting are such that equilibrium is not likely to be achieved in practice.

1. ZINC ALLOYS

Whilst there are two specifications, the vast majority of castings in the United Kingdom are made in alloy A, alloy B only being used where the greater hardness and strength are needed as, for example, in zip fasteners. A feature of both alloys is the tight control on impurities which is necessary to avoid premature failure due to intergranular corrosion, a phenomenon not uncommon in the early alloys.

TABLE 2.1. U.K. DIECASTING ALLOYS WHERE ONE VALUE IS QUOTED THIS IS THE MAXIMUM FIGURE.

Zinc base alloys BS 1004.

Alloy	Zn	Al	Cu	Mg	Fe
A	BAL	3.8 4.3	0.10	0.03 0.06	0.10
B	BAL	3.8 4.3	0.75 1.25	0.03 0.06	0.10

Ni	Pb	Cd	Sn	Tl	In
0.006	0.005	0.005	0.002	0.001	0.0005
0.006	0.005	0.005	0.002	0.001	0.0005

Aluminium base alloys BS 1490

Alloy	Al	Cu	Mg	Si	Fe
LM2	BAL	0.7 2.5	0.30	9.0 11.5	1.0
LM4	BAL	2.0 4.0	0.15	4.0 6.0	0.8
LM6	BAL	0.1	0.1	10.0 13.0	0.6
LM20	BAL	0.4	0.2	10.0 13.0	1.0
LM24	BAL	3.0 4.0	0.3	7.5 9.5	1.3

Mn	Ni	Zn	Pb	Sn	Ti
0.5	0.5	2.0	0.3	0.2	0.2
0.2 0.6	0.3	0.5	0.1	0.1	0.2
0.5	0.1	0.1	0.1	0.05	0.2
0.5	0.1	0.2	0.1	0.1	0.2
0.5	0.5	3.0	0.3	0.2	0.2

TABLE 2.1 *cont.*

Magnesium base alloys BS 2970

Alloy	Mg	Al	Zn	Mn	Cu	Si	Fe	Ni	Cu + Si + Fe + Ni
MAG 1	BAL	7.5 9.0	0.3 1.0	0.15 0.4	0.15	0.3	0.05	0.01	0.4
MAG 3	BAL	9.0 10.5	0.3 1.0	0.15 0.4	0.15	0.3	0.05	0.01	0.4
MAG 7	BAL	7.5 9.5	0.3 1.5	0.15 0.8	0.35	0.4	0.05	0.02	0.75

Copper base alloys BS 1400

Alloy	Cu	Zn	Sn	Pb	Fe	Al	Total impurities
PCB 1	57 60	BAL	0.5	0.5 2.5	0.3	0.5	0.5

Zinc has several advantages over its rivals. For example, its low melting-point, which leads to extended die life, its ability to be formed by cold working, and to be finished in an attractive way by chrome plating. This later feature has lead to extensive use for "decorative" castings. For large structural castings, the weight of zinc is a disadvantage, but this has been particularly overcome by the development of "thin wall" diecastings.

The melting of zinc poses no problems. In its simplest form, ingot is added at the machines, whilst at the other end of the scale hot metal is transferred to the machines by a launder system, ingot and returns being fed to a bulk melting furnace by a conveyor.

2. ALUMINIUM ALLOYS

Aluminium diecasting alloys contain silcon. The aluminium-silicon thermal equilibrium diagram (Fig 2.2) shows the alloy characteristics. The presence of other elements does not substantially alter this relationship.

The question of what constitutes a good diecasting alloy or which is the preferred alloy from the five listed is difficult to answer since there are several factors to be considered. An attempt to quantify

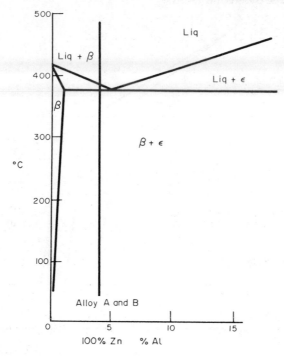

FIG. 2.1 THE Zn-Al SYSTEM

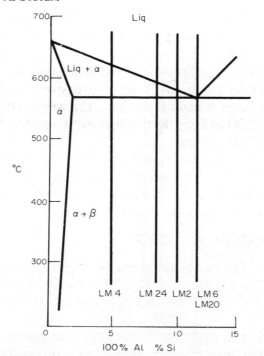

FIG. 2.2 THE Al-Si SYSTEM

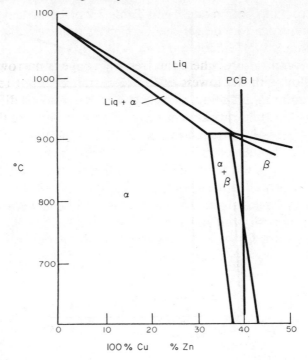

FIG. 2.3 THE Cu-Zn SYSTEM

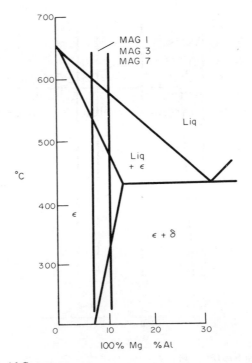

FIG. 2.4 THE Mg-Al SYSTEM

these factors has been made[1]* and Table 2.2 places the alloys in an approximate order for the various properties required. Fluidity is best in an alloy of high silicon content, which in this case is near to the eutectic composition, where the freezing range is narrow, and worst in the alloy with the lowest silicon. Resistance to hot tearing is not very different in any alloy, since if there were marked differences the worst alloys would quickly drop out of favour because this property is very important when using a rigid and complex die. Pressure tightness is another property where little difference is obvious. This feature is very much influenced by conditions in the die during freezing, but it would appear that the higher silicon, narrow freezing range alloys have a slight advantage. Pressure tightness is, not simply related to porosity but to its distribution, and in the wider freezing range alloys porosity may be higher but more favourably distributed.

Machinability may be important depending upon the eventual use, and here the lower silicon, high copper alloys have a distinct advantage. The poorest machinability is associated with a high silicon content and a low impurity level, giving the double disadvantage of high abrasion on the machine tool due to silicon particles and poor chip breaking action due to the lack of small intermetallics in the alloy matrix. Casting temperature is very similar for most alloys except the low silicon, high copper variant where the casting temperature could be substantially higher. A property not listed but of major importance is cost. Most diecasters purchase ingot or molten metal of the required analysis from secondary metal refiners who purchase metal from many other outlets. It follows, therefore, that the cheapest alloy will be the one permitting the highest impurity levels, since these alloys do not have to be refined or diluted with pure aluminium to the same extent.

Whilst the effect of the silicon level in an alloy is all important, the other additions or impurities have an important role to play.

Copper is the most common addition, and whilst it increases the hardness, its chief practical effect is upon machinability. Iron, whilst nominally present as an impurity is important in reducing the tendency of the alloy to "solder" to the die surface and in pressure diecasting it can be tolerated to a greater extent that in sand castings where it can give rise to embrittlement because of long brittle needles of iron-rich constituent.

In pressure diecastings these needles are much reduced and dispersed and have little effect on strength or ductility. The needle-like phase of

* Superscript numbers refer to References at end of chapter.

iron-rich constituent is influenced by manganese, and in sand or gravity alloys it plays an important role. It has, however, been said [1] that since the embrittling effect of iron is absent, manganese may do more harm than good. The use of, or rather the tolerance of, zinc was for many years limited due to its supposed effect of lowering corrosion resistance and causing hot shortness, and for many years was restricted in alloy specifications. Since much of the secondary material going for refining contained zinc, there was economic pressure to establish the level which could be tolerated and this is reflected in current specifications.

The melting of aluminium alloys for pressure diecasting causes few problems, but problems do exist.[1,2] Ingot is frequently added to melting/bale-out furnaces, or alternatively bulk melting furnaces are used and molten metal transferred. If the alloys are bulk melted, the temperatures are usually higher to allow for loss during metal movement to the bale-out units.

The higher temperature in bulk melting furnaces can help to prevent the formation of sludge in the alloys where this is a hazard. Sludge is caused by gravity segregation of iron- and manganese-rich constituents when the temperature of the alloy is relatively low (Table 2.3). In bale-out furnaces it can form a layer at the furnace bottom and the danger is that it can find its way into the castings. When such castings are machined, the segregates cause "hard spots" which can fracture tools and cause drag marks on a machined surface. The addition of ingot to a bale-out furnace encourages the formation of sludge, since the temperature of the melt adjacent to the ingot is lower. Once sludge has formed in a bale-out furnace it is very difficult to remove and the furnace is best emptied and the metal sent for remelting in a bulk unit. Sludge in a bale-out unit is prevented by keeping metal temperature above ~700°C and by regular stirring the bath, particularly if ingot is added.

3. COPPER ALLOYS

There is only one pressure diecasting alloy in the United Kingdom covered by BS 1400, and this is essentially a duplex brass. Figure 2.3 shows the copper-zinc thermal equilibrium diagram on which it is based.

The presence of lead, which is essentially insoluble in copper, is to aid machinability whilst aluminium prevents zinc oxidation and loss and improves fluidity.

Since the end use of many pressure diecast components is in the form of plumber's fittings for use in water, corrosion resistance is im-

TABLE 2.2 PROPERTIES OF ALUMINIUM PRESSURE DIECASTING ALLOYS

		Property			
Rank	Fluidity	Resistance to hot tearing	Pressure tightness	Machinability	Casting temperature
Best ↑	LM6, LM20	LM2, LM6, LM20	LM6, LM20	LM4, LM24, LM2	LM2
	LM2, LM24	LM4, LM24	LM2, LM4,	LM20, LM6	LM6, LM20, LM24
↓ Worst	LM4		LM24		LM4

TABLE 2.3 VARIATION IN METAL COMPOSITION BETWEEN TOP AND BOTTOM OF A MELT[1] OF LM24 ALLOY

	Cu	Mg	Si	Fe	Mn	Ni	Zn
Top	3.45	0.20	9.35	0.95	0.35	0.19	0.96
Bottom	2.96	0.07	10.10	5.00	2.00	0.64	0.61

portant. This problem can be solved in several ways. At elevated temperature the alloy has a duplex structure of alpha and beta phases, the percentage of beta becoming less as the temperature is lowered. Beta-phase is necessary at high temperature to give resistance to hot cracking, but its presence at room temperature can lead to "dezincification" which is the corrosion mechanism whereby the alloy is degraded. It follows that if the alloy has a very carefully controlled zinc content, there will be beta-phase at high temperature but little at room temperature and at room temperature the beta phase will be in an isolated condition and corrosion will not proceed to any great extent. Corrosion of the alpha phase can be inhibited by small amounts of arsenic.

Another approach to the problem is to give the castings a low-temperature heat treatment to achieve the same result as just described, whilst a further answer is to introduce alloying elements. There are several proprietary alloys which achieve this, generally by the addition of manganese, nickel or tin.

The high casting temperature of brass means that die life is drastically reduced as compared to the other metals, and this has lead to the development of die materials based on molybdenum and tungsten.

4. MAGNESIUM ALLOYS

The composition of the present diecast alloys is covered by BS 2970 and of the three listed only MAG7 is cast to any extent in the United Kingdom. Magnesium alloys find application where weight is important, in such articles as portable saws, lawnmower housings, cooling fan housings, automobile engine components, television camera tripods, etc. The corrosion rating of magnesium alloys is low, and in practice components are either painted or chromate treated since MAG7 is not easily plated like zinc nor will it retain a high lustre when polished. Melting is often considered a drawback, but there are no difficulties which cannot be overcome. Beryllium additions up to ~ 0.001% help to combat oxidation, but even so, melting has to be carried out under a protective atmosphere, usually SO_2 gas, but other gases have been tried. The melting furnace, has to have a cover with access doors through which ingot can be charged and molten metal ladled out or the gooseneck of a hot chamber machine inserted (Fig. 2.5[3] and 2.6[3]).

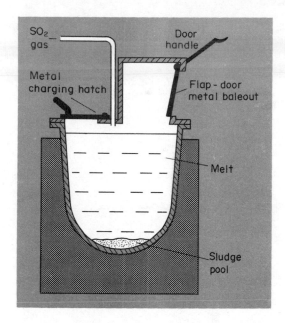

FIG. 2.5 DIAGRAM OF SIMPLE HOLDING-CRUCIBLE COVER
WITH DOORS FOR CHARGING AND BALE-OUT, AND
SINGLE GAS SUPPLY PORT.

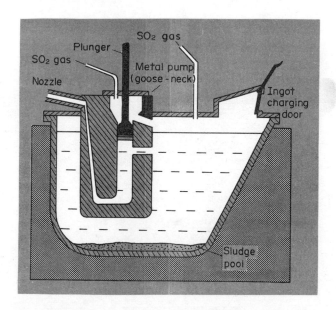

FIG. 2.6 DIAGRAM OF SIMPLE COVER ARRANGEMENT FOR
HOT-CHAMBER CRUCIBLE, WITH PROTECTIVE
GAS (SO_2) SUPPLY TO MELT CHAMBER AND TO THE
GOOSENECK TOP.

REFERENCES

1. F. H. Smith, Aluminium alloys for pressure diecasting, *Metal Industry,* 1960, vol. 7, pp. 127-129, vol. 8, pp. 147-149.
2. E. G. Morgan, Diecasting Metal Control. *Foundry Trade Journal,* Oct. 22, 1981. pp. 676-681.
3. T. B. Hill, *BNF Guide to Better Magnesium Diecasting.* BNF Metals Technology Centre, Grove Laboratories, Denchworth Road, Wantage, Oxon.

Chapter 3

Pressure Diecasting Machines

A diecasting machine is a complex machine tool having mechanical, electrical, hydraulic and pneumatic features. It is built to high engineering standards and is expected to operate for many years in a hostile environment with a minimum of maintainance and attention. It looks a complicated confusing piece of equipment, and this is because machine developments over the years have been "added on" to a basic simple design. This "adding on" process has, in the interests of maintainance and general accessibility, been external to the machine with the result that a modern machine has become a maze of pipes, valves and switches (Fig. 3.1). The type of control panel on a modern machine can also appear very complex (Fig. 3.2).

1. GENERAL MACHINE CHARACTERISTICS

A modern machine is best understood from the basic machine design. Details of individual machine designs can vary considerably, but the basics are as shown in Fig. 3.3. A machine has a bed (G) on which is mounted a fixed platen (C) and a thrust platen (A). Joining the platen are four tie bars (D). On the fixed platen (C) is mounted the injection system (H), and on the thrust platen is mounted the locking ram (F). Between the fixed and thrust platens is a moving platen (B) which slides on the bed of the machine and is guided by the tie bars. The moving platen is attached to the thrust platen and the locking mechanism by mechanical levers called toggles (E), which cause the moving platen to move as required. The die halves (J) are mounted on the fixed and moving platens and the machine is "locked" when the die halves are pressed together by the action of the locking cylinder and toggle mechanism. To accommodate dies of different sizes and to allow for expansion due to temperature, the lock on a machine can be adjused either by large nuts on each tie bar or by a central adjustment device. The load on the die is transmitted via the toggles to the fixed and thrust platen and hence to the four tie

Fig. 3.1 B Series Buhler Diecasting Machine

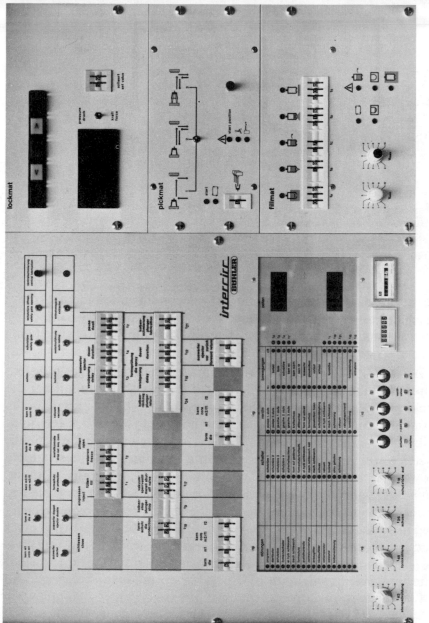

Fig. 3.2 Part of the Control Panel of a B Series Buhler Diecasting Machine, Showing Additional Panels for Control of Locking Force (lockmat). Autoladle (fillmat) and Casting Extractor Unit (pickmat).

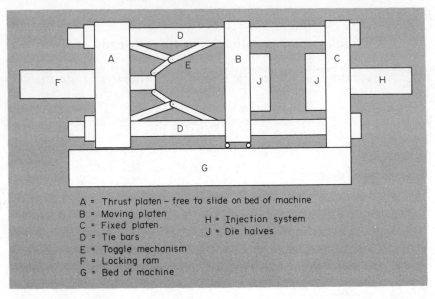

A = Thrust platen – free to slide on bed of machine
B = Moving platen
C = Fixed platen.
D = Tie bars
E = Toggle mechanism
F = Locking ram
G = Bed of machine
H = Injection system
J = Die halves

FIG. 3.3 DIAGRAMMATIC REPRESENTATION OF THE MAIN COMPONENTS IN A DIE-CASTING MACHINE

bars. The load on the tie bars can vary depending upon the size of the machine and in the United Kingdom most machines carry from ~ 50 to 800 tons load, and this is the most common way of specifiying machine size. The other method is by quoting the platen dimensions.

After locking the die, a casting is made when the injection system forces metal into the die cavity. After the metal has solidified, the locking mechanism moves the moving platen backwards and the casting is removed from the die. The whole sequence is then repeated.

It follows that there are only two hydraulic or pneumatic circuit systems which are fundamental to the machine, and these supply power to the locking cylinder and to the injection cylinder. However, lubrication is necessary for the moving parts, which in practice means lubrication pipes to the toggle pins, moving platen/ bed interface and the injection cylinder. No doubt the early machines had no more than the above, and even today it is the pressures developed in the systems relative to movement of the locking ram and injection piston which are of crucial importance in diecasting. All other systems which cause the machine to appear so complex are basically ancillary systems installed for reasons of convenience, safety, increased die complexity, etc. Typical systems are as follows.

Primary Systems

1. A hydraulic or pneumatic circuit to both ends of the locking cylinder.
2. A hydraulic or pneumatic circuit to both ends of the injection cylinder.

Both the above have facilities for control of pressure and speed of movement.

Secondary Systems

1. A hydraulic or pneumatic circuit to operate the casting ejector mechanism, mounted on the back of the moving platen.
2. A hydraulic or pneumatic circuit to operate core-pulling devices for more complex dies.

Safety Systems

These are generally connected with the movement of guards which protect the operative from possible metal splashing if the die "flashes" or to prevent access to the machine whilst parts are in motion. They can also prevent involuntary movement. Safety systems frequently operate by preventing other functions on the machine from taking place unless the safety circuits are in the closed position indicating that all guards, etc., are in place.

Lubrication Systems

These are similar to those found on any mechanical device and they supply lubricant to toggle pins and bushes, machine sliding bed, shot sleeve and plunger, etc.

Optional Systems

These can vary, but typically are related to the following types of operation.

1. Automatic die spray
2. Automatic castings removal
3. Automatic casting detection
4. Automatic molten metal transfer

The above is not necessarily a complete list, but a modern machine can easily have ~10 hydraulic or pneumatic systems, each with its associated electrical circuits, valve work, hydraulic pipes, etc.

2. MEASUREMENT OF MACHINE PERFORMANCE

Machine performance checking is not undertaken for academic interest, but in order to establish whether the machine has the required locking force to hold the die halves together and whether the injection system is capable of making the castings required. A new machine should be checked initially to see if it complies with the maker's specification, whilst continual checks should be carried out to see if or how the performance in service is deteriorating to the point at which an unacceptably high casting reject rate is being produced.

In practice, measurements are made on the locking system to measure locking force and locking cylinder pressure, whilst on the injection system measurements are made of metal injection velocity and the pressure developed in the injection cylinder associated with these velocities.

Generally speaking the above are the only areas of interest. With the other systems it is frequently self-evident that they are operating satisfactorily or otherwise.

The Injection System

FUNCTION AND ACTION OF THE INJECTION SYSTEM

The performance of the injection system is fundamental to the production of good quality diecastings, controlling the metal pressure during and after cavity fill as well as the volume of metal and the velocity with which it is injected into the die.

The subject of injection system performance and efficiency is now receiving much attention at the academic level[1-4]* and it is important that this work and associated work on die design be broadly understood by technicians and applied on the shop floor.

A knowledge of the injection performance of all the machines in a diecasting foundry enables the machines to be used more efficiently. Die running and gating systems can be designed with greater precision and there is more certainty that dies will be right first time and thus expensive trial-and-error methods can largely be avoided.

If the die requirements with regard to metal injection velocity and pressure are known, the die can be matched at the production planning stage to machines capable of giving these requirements[3] (Fig. 3.4) rather than by trial-and-error methods on the shop floor. New

*Superscript numbers refer to References at end of chapter

machines can be purchased on the more rational basis of injection characteristics and different manufacturer's machines can be more objectively compared. It is also easy for any foundry to see where there are gaps in their existing machine range.

FIG. 3.4   BNF GATEWAY DIE DESIGN SYSTEM

Whilst the function of any injection system for any metal being cast is essentially one of introducing metal into the die cavity, there are fundamental differences in achieving this depending upon the alloy involved.

The components of a diecasting machine and injection system are made of cast iron and steel, so if the alloy being cast is aggressive to these materials at the casting temperature, contact between the two must be minimised. Of the two commonly diecast metals, aluminium is aggressive, whilst zinc is not at the temperatures involved. This has lead to a divergence in machine design, zinc being cast by the hot chamber process and aluminium by the cold chamber process. Of the less common metals in the United Kingdom, brass must also be cast by the cold chamber process, whilst with magnesium either process can be used. The words "hot" and "cold" are, of course, only relative terms used to describe the processes.

Figures 3.5 and 3.6 show the essential features of each type of machine. The cold chamber process (Fig 3.5) is simply a cylinder mounted onto the fixed platen. At one end, the cylinder protrudes through the platen into the running system of the die cavity, whilst in the other end a plunger is activated by the injection system. At the upper side of the cylinder is the hole through which metal is poured.

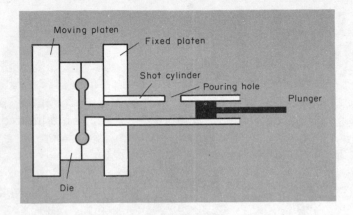

FIG. 3.5 DIAGRAM OF THE COLD CHAMBER PROCESS

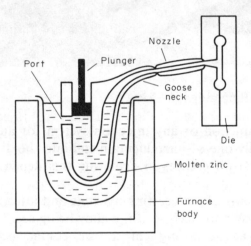

FIG. 3.6 DIAGRAM OF THE HOT CHAMBER PROCESS

A casting is made when the plunger advances down the shot cylinder, covering up the pouring hole in the process, and the metal is injected into the die cavity. After freezing, the die is opened, the casting removed and the plunger is returned to its previous position. All diecasting machines can be fitted with different diameter shot cylinders to accommodate various casting weights and for varying the pressure on the metal. Likewise the shot cylinder can be mounted in a variety of positions on the fixed platen to accommodate different die sizes and running systems. The large majority of cold chamber machines have their injection system mounted in the horizontal plane, but there are some where it is mounted vertically.

The hot chamber process is shown in Fig. 3.6 where it is seen that the lower part of the injection system is contained in a bath of zinc maintained at casting temperature by a furnace and burner system. This part of the system is called the gooseneck, and it is connected at one end to the die by a nozzle and at the other end to a plunger attached to the injection system. As in the cold chamber process, the injection plunger diameter can be varied to cater for different shot weights and to give different metal pressures. Metal is admitted to the gooseneck through a port and a casting is made when the plunger advances down the shot sleeve, covering the port in the process. The metal is then injected via the gooseneck and nozzle into the die cavity. After freezing, the die is opened, the casting removed and the plunger is returned to the original position. As the plunger is returned it uncovers the port in the gooseneck and molten metal is admitted; this constitutes the next shot. Hot chamber machines have their injection systems mounted vertically or at an angle, and metal is replenished in the pot either by molten zinc from a bulk melting furnace or by ingot.

The injection system of either process can be complex and can differ in many respects, depending upon the process and whether the power is hydraulic or pneumatic. The simplest system is on the hot chamber pneumatic machine (Fig. 3.7). Typically air pressure of ~100 lb/in^2 is constantly applied to the underside of the injection piston to keep it in the return position, the top of the piston being subject only to atmospheric pressure. When a casting is made the same air pressure (~100 lb/in^2) is applied to the top of the piston by means of a valve which simultaneously closes the cylinder to atmospheric pressure, and because of the difference in area over which the same air pressure is applied there is a downward force transmitted to the metal to make the casting. During injection the piston does not move at constant speed. There is a slow first stage, during which the metal is being moved through the gooseneck, nozzle and part of the die runner system as far as the gates at the actual casting. At this point a trip switch on the machine, which activates a valve, causes the piston to increase velocity (second stage) and actually fill the casting cavity at the higher velocity required for good quality castings. The actual velocity obtained in the first and second stages is controlled by valves on the machine which can be varied according to requirements. To avoid the build-up of excessive back pressure due to compression of air on the underside of the piston and hence a reduction in force applied to the metal, there is a valve which opens during the injection phase of the cycle. When the casting has solidified, the valve under the piston is closed, thus allowing the pressure to build up to ~100 lb/in^2, whilst the valve on top of the piston is vented to atmosphere and hence there is an upward force on the piston which causes it to return to its original position ready for the next shot. In practice the air pressure is usually obtained from a reservoir tank at the side of each machine which is charged up to pressure, generally by air from a

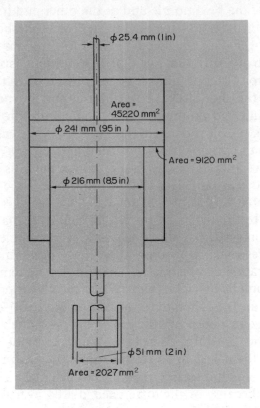

Fig. 3.7 Schematic Arrangement of Injection Cylinder and Plunger for a Hot Chamber Machine

ring main system. This method gives a minimum pressure drop during injection, since the reservoir volume is substantially greater than the injection cylinder volume. If a machine were connected directly to an air ring main system, air pressure in the injection cylinder and hence the force could fall during injection because the small diameter pipes of the ring main could not supply air at the required rate. This problem could be particularly acute when several machines were operated from the same ring main.

The most complex injection systems are found on hydraulically operated cold chamber machines and these can vary considerably in their detailed design and working mode, but in principle the method of working is typically along the following lines. Figure 3.8 shows a hydraulic injection cylinder and shot cylinder. Pressure is applied to the head side of the piston, which causes the piston to move the metal in the shot cylinder and make the casting. Pressure on the annulus side of the piston is avoided by the opening of a valve, which allows the hydraulic fluid to return to the storage tank. Hydraulic fluid is at the annulus side in order to return the piston to the original position after making a casting. The movement of the piston during injection is not at constant speed, but takes place in three stages, the changeover points being controlled either by electrical trip or proximaty switches, by timers or displacement transducers on the machine or by the build-up of hydraulic pressure in the system activating valves of various types. The actual first and second stage velocity can be varied by valves in the hydraulic system, which usually work by restricting the flow of hydraulic fluid. The first injection stage, or slow approach, is designed to start the movement of metal in the shot cylinder in such a way as to avoid the formation of waves and to expel as much air as possible from the cylinder and die cavity.

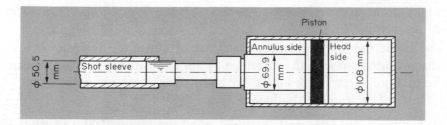

FIG. 3.8 SCHEMATIC ARRANGEMENT OF A COLD CHAMBER INJECTION UNIT

The shot cylinder is commonly filled to ~60% of its volume, the remainder, being air which if not expelled through the die, could become entrapped, and give rise to casting defects. The first stage continues at a relatively slow constant speed until the metal reaches the gates between the running system and the die. At this point the fast phase or second stage is actuated, which forces metal into the die cavity at the appropriate speed for the casting being made. When the cavity is full and the metal has partially frozen so as to have a solid outer layer, a third phase known as the intensification phase is applied. Not all machines have this feature, whose purpose is to apply a high pressure to the casting during the final stages of solidification to compensate for shrinkage and gas porosity and minimise the occurrence of defects due to this phenomenon. During the intensification phase there is very little piston movement, only that necessary to compensate for shrinkage and gas porosity. After a casting has been made, the plunger is returned to the starting position by a build-up of pressure on the annulus side of the piston in the injection cylinder.

During the second and third phases of injection the hydraulic pressure is supplied from a reservoir called an accumulator (the first phase is usually driven from the pump). An accumulator is a device for storing energy built up over a long period for release over a short period. It generally consists of a steel cylinder containing nitrogen gas which is compressed by oil from the hydraulic pump. The oil and nitrogen are separated by a piston. By compressing the nitrogen, energy is stored and this is used to drive the injection system. The presence of an accumulator enables a much smaller pump to be employed than would be needed if the pump were to provide the power directly to the injection system. The actual second stage of the injection cycle, when force is needed quickly, may only be a small percentage of the total casting cycle and hence a large pump would be uneconomic. A small pump, however, can keep charging the accumulator during the whole of the cycle ready for a release of energy when required. It follows from the above that the nitrogen pressure and volume are important parameters in the injection performance of a machine.

Despite the fact that many machines do not have any intensification or third phase in the injection cycle, intensification pressure can be a major factor in reducing both shrinkage and gas porosity.

Intensifiers can be obtained with control over:

1. the degree of intensification, called the intensification ratio;
2. the rate of intensification,
3. the delay between the end of injection and the start of intensification. These variations are shown in Fig. 3.9[5]

TYPES OF INTENSIFICATION SYSTEMS

On modern machines, two types of system are in common use, the prefill and the hydraulic intensifier.

On some old machines the pressure is simply increased by bringing in a high-pressure pump at the end of the injection stroke, with no further sophistication.

Prefill System

Prefill systems employ a cylinder within a cylinder as shown in Fig. 3.10. The inner cylinder is actually a hollow piston which receives hydraulic oil directly from the accumulator, producing stage 2 plunger movement. The outer cylinder which is responsible for inten-

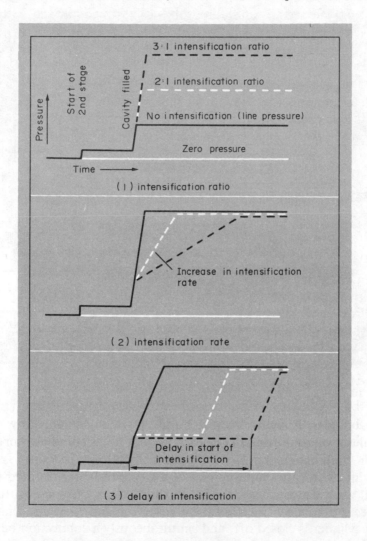

Fig. 3.9 Variables in Intensification Systems

sification receives oil from an atmospheric reservoir during the injection stroke and from the accumulator during the intensification stage of the cycle. The important part of the system is the response and reliability of the valve opening the outer cylinder to the accumulator.

During the fast injection stroke (stage 2) accumulator oils flows to the inner piston. At the same time oil drawn through the prefill valve from the atmospheric reservoir maintains the outer cylinder filled with oil behind the large diameter end of the piston. When the die is filled, the resistance to motion causes the pressure to build up in the

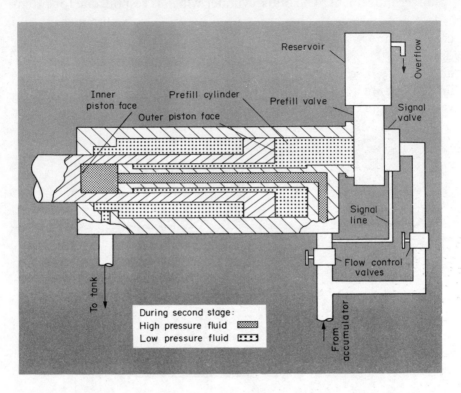

FIG. 3.10 PREFILL INTENSIFICATION SYSTEM

inner cylinder. It is this pressure build-up which actuates flow of accumulator oil to the outer cyclinder, so that the accumulator pressure acts in both chambers, providing intensification. Actuation is complex: inner pressure build-up causes a signal to be transmitted to the prefill valve which causes a spool to shift. When the spool shifts, the connection between the prefill oil reservoir and the large diameter outer cylinder is closed off, and simultaneously a connection between accumulator and outer cylinder is opened. The pressure therefore acts both in the inner and outer cylinder, thus producing an intensified pressure on the metal.

The system can incorporate control over the intensification rate and the delay between the end of injection and the start of intensification. It could, by incorporating a different pressure system to the inner and outer cylinders, also provide an adjustable intensification ratio.

Hydraulic Pressure Intensifier

In these systems the increase in metal pressure is obtained by increasing the hydraulic pressure within the injecton cylinder. This is done by using two hydraulic cylinders, the main injection cylinder and an intensification cylinder—sometimes called a multiplier. The intensification cylinder can be mounted in-line (Fig. 3.11), parallel or vertical to the main injection cylinder.

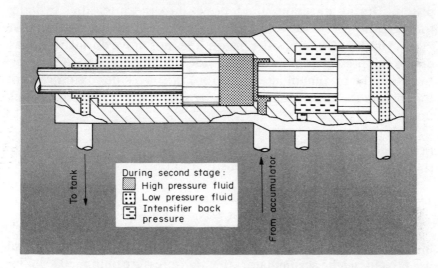

During second stage :
☐ High pressure fluid
☐ Low pressure fluid
☐ Intensifier back pressure

To tank

From accumulator

FIG. 3.11 HYDRAULIC PRESSURE INTENSIFIER (MULTIPLIER)

During the second (fast) stage of injection, oil is supplied from the accumulator to the main cylinder. The intensifier cylinder can be brought into operation from a signal generated by a limit switch (sometimes the same switch as used to initiate the fast injection stage), or by the pressure rise at the end of injection. In some cases, unless care is taken in setting the controls, the intensifier piston can bottom, resulting in no intensification. A check valve prevents flow from the main cylinder so that at the end of the stroke the main cylinder is sealed off and the pressure is intensified by the resultant force from the intensifier cylinder, produced by the pressure on the head side and annulus side (back pressure). The maximum intensification ratio is governed by the ratio of the area on the intensifier piston. For diecasting machines this ratio is usually between 1.6:1 and 4:1.

This system can be designed so that the intensification ratio, rate and delay can be varied. The ratio can be varied by controlling the pressure on the annulus side of the intensifier piston by using a regulating valve.

The final pressure can be controlled independently from the line pressure to the main injection cylinder by using a separate pressure supply and accumulator in the intensifier circuit. The intensification rate can be varied by incorporating a flow control on the inlet side or annulus side of the intensifier. The delay in the start of intensification is obtained by delaying the supply to the intensifier or by release of the intensifier back pressure.

This type of system can incorporate a check valve in the intensifier piston. During injection, fluid flows into the intensifier cylinder, through the check valve inside the intensifier piston, and into the main cylinder. The intensifier piston does not move during this phase due to the pressure on the annulus side. If this pressure is set to zero to obtain the maximum intensification, the intensifier piston can bottom before the end of the injection stroke resulting in no intensification.

With the internal check valve system the supply used during injection is also used during intensification. Therefore, slower injection speeds also result in slower intensification.

High Pressure Pump

This system was used on some machines produced up to the mid-1960s. A low-pressure pump of approximately 1000 lb/in^2 was used to charge the accumulator; the high-pressure pump, 2000 lb/in^2, was brought in at the end of injection to produce intensification. Because the intensification is produced from a pump rather than an accumulator, the rate of pressure build-up with this system is rather slow.

MEASUREMENT OF INJECTION PARAMETERS

Instrumentation

To measure machine performance it is necessary to have the proper equipment, which will not only do the job satisfactorily but will be capable of working in a relatively hostile environment. The following are the requirements and the parameters to be measured.

1. To be able to measure pressures up to the maximum likely to be developed, making full allowance for inertia effects, and to

accurately estimate the points at which pressure changes are occurring.

2. To be able to measure dry shot piston velocity and the velocity of metal movement in the shot cylinder, up to 25-30 ft/sec and to accurately estimate the points at which changes of velocity are occurring.

3. We need to measure several parameters simultaneously, the most common being velocity of plunger movement and the pressures either side of the injection piston.

In practice, because of the need for rapid response to changing parameters and to eliminate as far as possible inertia effects, the most commonly used equipment is based on the multi-channel oscilloscope or ultraviolet recorder coupled to pressure and displacement transducers whose outputs consist of electrical signals.

Many diecasting machine manufacturers market an instrument package, as do some instrument companies and other organisations, but there is no reason why foundries should not build their own from individual components. Methods of measurement can vary, particularly with regard to measuring the piston velocity and also whether the equipment is mounted semi-permanently on the machine or is intended to be quickly transferable from one machine to another. There is probably no right or wrong way when it comes to instrumentation; it is more likely a personal preference or with what one becomes familiar.

A typical set of instruments is shown in Fig. 3.12. The instrument package is a deliberately simplified version of the equipment used in research work and has, with only minor modifications, been used in foundries for many years. The system consists of either a six- or twelve-channel ultraviolet (U.V.) recorder with suitable amplification. The amplifiers are non-adjustable and hence cannot be accidentally moved when in the foundry.

Normally, two pressure transducers are supplied either to cover the range $0 - 300$ lb/in^2 for pneumatic machines or $0 - 5000$ lb/in^2 for hydraulic machines; thus the system can measure impact peaks if present. The system is such that only two pressure transducer leads are necessary, which then combined with a switch to a "high" or "low" position will give the two ranges $0 - 300$ lb/in^2 and $0 - 5000$ lb/in^2 as full-scale measurements over the 150 mm wide scale on the U.V. instrument. The system is calibrated electronically and there is generally no need to calibrate the transducer pressure output measurements on the U.V. trace against pressure on a standard pressure gauge. Measurement of piston movement is achieved using a displacement transducer having a wire length of 36in. Again there can be more than one setting for full scale to improve accuracy for short piston strokes. The transducer is a helical potentiometer and does not measure velocity directly. Velocity is measured on the U.V. paper indirectly by

FIG. 3.12 BNF METALS TECHNOLOGY CENTRE, DIECASTING DIAGNOSTIC TROLLEY

measuring distance obtained from the displacement transducer and by measuring time which is obtained from the paper speed setting on the U.V. instrument. Figures 3.13 and 3.14 show typical pressure and displacement transducer mountings on a cold chamber diecasting machine. Pressure transducers fit into holes normally occupied by

FIG. 3.13 Pressure Transducer Mounted at the Shot End of a Machine

FIG. 3.14 Displacement Transducer and Extension Arm on Injection
Plunger Ram

blanking plugs, but there is no reason, if speed is essential, why they should not be connected to the hydraulic circuit by quick-release fittings. The displacement transducer is generally fastened to a plate attached to the casing on the injection cylinder. Normally only two bolts and wing nuts are needed for fastening and for quick release. The wire on the displacement transducer is attached to a short stiff rod fastened to the moving ram and is thus free to move in and out of the transducer cover under the action of the ram and a return spring inside the transducer. With hot chamber machines mounting of instruments is basically the same, but with pneumatic machines the pressure transducers should be as close to the injection cylinder as possible to minimise errors. Similarly the displacement transducer should not be subjected to temperatures over ~ 200°C. This temperature could be reached in certain areas of the injection system on a hot chamber machine, so it is best to mount the displacement transducer on a small specially built subframe fastened to the top of the injection cylinder; the moving wire can then be fastened to the tail rod which passes through the cylinder end cover. If there is no tail rod there is little choice but to mount an arm on the injection rod as in the cold chamber process and connect the displacement transducer wire to this rod by an extension wire. The disadvantage of this method is the lack of rigidity in the arm, which can cause excessive oscillation and give false readings on the displacement trace.

Connecting the pressure and displacement transducers to the U.V. recorder is ~15 - 20 ft of cable to enable the U.V. recorder to be operated at a suitable distance from the machine under test. The system is completed by connecting the U.V. recorder to the main electric supply by a similar length of cable. The U.V. instrument has several control knobs, switches and levers, but fortunately in diecastings most are never changed once set. The controls are frequently as follows.

Mains On/Off Switch

This is switched on at the start of the exercise and off when it is finished. Sometimes this switch also controls the lamp, but sometimes a separate switch is provided for this depending on the make of U.V. recorder.

Trace Intensity

This usually has two levers or knobs, one to control the intensity or degree of contrast between the transducer trace and the paper, the other to control the intensity or contrast between the timing line traces and the paper. The degree of contrast necessary should be found by experiment, but once established the controls need not again be adjusted.

Timing Lines

The timing line controls are usually a series of push buttons, only one being pressed in as required. The most common timing line for diecasting is the 0.1 sec interval line. Timing lines are lines running across the paper and their distance apart is dependent on paper speed. They are most useful when measurements are being taken on the traces and they serve as a reminder of the paper speed being used. If the 0.1 sec interval line only is needed, there is no need to consider the other control buttons once it is set.

Event Marker

This is a push button which when pressed causes a small beam of light to mark the edge of the U.V. paper. It is not normally used in diecasting.

Alternative Scale for Transducers

This is not normally found on the instrument, but is an addition for diecasting purposes, if it is necessary to use transducers of different outputs, etc. The control is usually a simple switch having "high" and "low" positions or a rotary switch if more alternatives are needed.

Paper Speed Control

Paper speeds are usually controlled by push button switches and for diecasting purposes the most usual settings are betwen 100 and 1000 mm/sec.

Instrument On/Off Switch

This is either a push or toggle switch and it is pressed to start paper movement and pressed again to stop. Having set the other controls it is frequently the only control used when actually checking a machine in the foundry.

Galvanometer Adjustment

This adjustment is not carried out at the front of the U.V. instrument but via a small access door on the top. A separate galvanometer is needed for each trace and the position of each trace on the paper is governed by the galvanometer adjustment. This is a useful feature since, when several traces are recorded on the same paper, confusion can be avoided by having starting zero positions set at different distances across the width of the paper.

Calibration of Instruments

Not all instruments are calibrated electronically before delivery, and even those which are calibrated should be checked occasionally, particularly if new transducers are fitted.

Pressure Transducers

Most diecasting machines have a pressure gauge fitted, and this can be used to calibrate the transducers. Pressure in the hydraulic system should be increased in increments of ~ 20 - 5000 lb/in^2 depending upon the maximum pressure in the system and the paper should be set on a fairly slow speed. When the exercise is completed, a length of paper 6 - 12 in long will be obtained with the differences in pressure shown as steps on the trace. This can be used to construct a graph of pressure against the trace displacement and this graph is used to obtain all subsequent pressure values from the trace measurements.

Displacement Transducer

This is calibrated against a normal steel rule. The paper is set on a fairly slow speed and the trace wire is extended in increments until it is fully extended. As before, the trace shows a series of steps and these values can be used to construct a graph of distance moved against the distance moved on the trace. These trace distances are then used to calculate velocity. Velocity is measured indirectly by measuring time and distance.

There are two ways of measuring velocity in practice. The first method is simple, rapid but possibly slightly less accurate, and involves the measurement of the angle of the trace using a protractor. The protractor angle is then compared to a graph of angle against velocity previously constructed. This graph is constructed as follows:

1. Set the paper speed to 100 mm/sec and timing marks to 0.1 sec and run out ~24 in of paper. Tear off from the roll. This will give a length of paper with no trace marks, but there will be timing marks across it at 0.1-sec intervals.

2. From the relationship between actual displacement movement against trace movement obtained from the previous graph or from the electronic calibration, select a suitable distance (usually 1 m or 1 ft) and mark the equivalent measurement to this distance vertically up one of the timing lines near the edge of the paper. Draw a vertical line up to this height, e.g. if the electronic calibration is given as 150 mm on trace = 30 in actual, we have 60 mm on trace = 1 ft actual. Therefore the vertical distance to be marked is 60 mm.

3. From the timing line chosen, count back ten divisions to give 1 sec in time, and draw a line joining the two timing lines.

4. A horizontal line representing 1 sec and a vertical line representing 1 ft is available. Therefore, if the end of the 1 sec line is joined to the end of the 1 ft line, a line which makes an angle to the horizontal corresponding to a velocity of 1 ft/sec is produced.

5. Measure the angle with a protractor.

6. Repeat the above procedure for as many velocities as one requires, e.g. If the 1 ft vertical line is retained and 0.5 sec is marked, the angle is 0.5 sec = 1 ft or a speed of 2 ft/sec. Similarly a mark of 0.25 sec and 1 ft vertical = 4 ft/sec and 0.10 sec and 1 ft vertical = 10 ft/sec.

   The protractor angles can then be tabulated against velocity (Table 3.1) and a graph constructed as shown in Fig. 3.15.

7. This graph is the master graph, and when real traces are examined, only the angle which the trace makes with the horizontal is measured. The injection velocity is then obtained from the graph.

8. Repeat stages 1 - 7 at different paper speeds to obtain a series of graphs at the paper speeds of interest.

TABLE 3.1 PROTRACTOR ANGLE VS. VELOCITY WITH PAPER AT 100 MM/SEC

Angle°	Velocity ft/sec
31	1.0
50	2.0
63	3.35
71.5	5.00
76	6.65
80	10.00

The second method of measuring velocity is the direct method of measuring distance and time on each trace.

For example, suppose the trace distances are measured and these correspond to real figures of 3.82 in and 0.048 sec, a velocity of $\dfrac{3.82}{12} \times \dfrac{1}{0.048} = 6.63$ ft/sec is obtained.

This method is probably more accurate than the measurement of angles, since if the angle were measured to ~1 degree of accuracy the 6.63 ft/sec would be between 75° and 76° on the graph shown in Fig. 3.15. If the angle method of velocity is used, therefore, accuracies of ~ ½ degree are needed, and this becomes more important at the higher velocities.

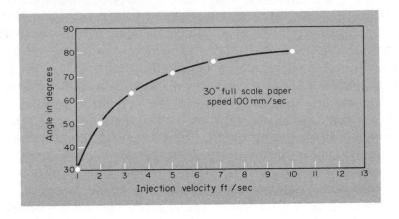

FIG. 3.15 GRAPH OF INJECTION VELOCITY VS TRACE ANGLE

Learning About the Instruments in the Foundry

Operating instruments under foundry conditions is not difficult, but a foundry is frequently a dirty, hot, noisy environment, communication may be difficult and under these conditions even experienced personnel make mistakes. It is, therefore, essential to be thoroughly familiar with the control and operation of the U.V. recorder, otherwise uncertainties will only add further confusion. It is best if the familiarisation process is done in a less stressful environment than the diecasting foundry. Once familiar with the mechanics of the equipment, traces should be made on a machine at various paper speeds to gain a "feel" for the instruments and for what is being recorded. There are no right or wrong ways to gain experience, and the beginner should be prepared to expend much paper in experimentation. Generally speaking if the U.V. recorder is set up to record from two pressure and one displacement transducer it is suggested that the pressure transducers be disconnected and readings initially taken from the displacement transducer only.

This is because events tend to happen more slowly with piston movement than they do with pressure changes. Examination of the traces will readily show the first and second injection stages as lines having different gradients with a definite changeover point between the two (Fig. 3.16) and, if the paper speed is fast enough, the third phase of injection may possibly be detected at the end of the second phase.

The next step is to connect the transducer on the pressure side of the injection piston and take traces in conjuction with the displacement traces. If an intensifier is fitted to the machine it should be in the "off" position.

The pressure trace will show a slight initial rise as the piston starts to move, followed by a steady condition during the first stage of injection. At the changeover to the second stage, or a short time after, the pressure will rise rapidly as metal is forced through the die gating system. At the end of the casting sequence the pressure will level off (Fig. 3.17).

The experiment should be repeated, and the effect of varying the injection velocity and pressure controls can be seen on the traces by different gradients on the displacement traces, and different peak positions on the pressure traces. The intensifier should now be brought "on" and the traces will show how variations in the intensifier control system will effect the maximum pressure reached, the delay in pressure application and rate of pressure build-up (Fig. 3.18).

Finally the pressure on the annulus side of the piston should be recorded in conjunction with the other traces (Fig. 3.19).

During this initial experimentation period with the instruments it is important to develop habits of working so they become automatic. All traces should be numbered in sequence in the same place on each trace. The reason for numbering is obvious, but numbering in the same place is a safeguard against accidentally examining a trace upside down and arriving at some strange conclusions! Similarly, details of the control settings should be recorded on a separate sheet, alongside the trace number. Avoid writing on the trace at this stage, but, if it has to be done, it should be done in pencil which can be erased if necessary, so as not to confuse the later measurements, which must be done on the traces.

Remember that the U.V. paper is sensitive to light and the trace will eventually fade, so traces should be kept in a box with a lid.

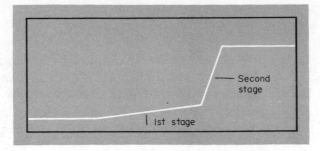

FIG. 3.16 DISPLACEMENT TRANSDUCER TRACE SHOWING 1st AND 2nd
INJECTION STAGES

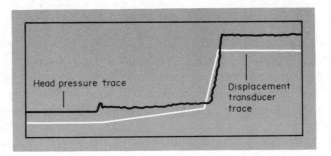

FIG. 3.17 DISPLACEMENT TRANSDUCER TRACE AND PRESSURE TRACE ON HEAD
SIDE OF INJECTION PISTON

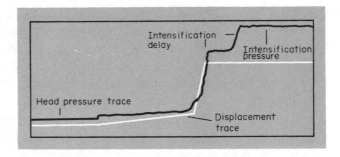

FIG. 3.18 AS FIG. 3.17, BUT SHOWING INTENSIFICATION DELAY AND PRESSURE

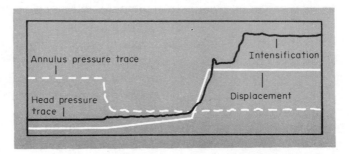

FIG. 3.19 AS FIG. 3.18, BUT SHOWING ANNULUS SIDE PRESSURE

Measurements on Traces, and Calculations

Pressure Measurements

Pressure measurements on the metal in the die cavity are very difficult to make, but an estimate is possible by measuring the forces acting upon the metal in the shot cylinder. The pressure on the metal at the end of the cavity-fill can be obtained and the forces calculated by substracting the annulus force from the head force and dividing by the shot plunger area, i.e.

$$\text{Metal pressure} = \frac{(\text{head pressure} \times \text{head area}) - (\text{annulus pressure} \times \text{annulus area})}{\text{shot plunger area}}$$

For example, the following sequence of measurements and calculations can be taken on an injection system as shown in Fig. 3.8.

1. Measure height of the trace from the zero position to the cavity-fill position for the head side pressure transducer.
2. Convert this measurement into pressure from the previously obtained calibration chart or graph (Assume it to be 1800 lb/in^2).
3. Repeat 1 and 2 for the annulus side (Assume it to be 1800 lb/in^2).
4. Calculate force on head side of piston.
   Force on head side = head pressure × head area

   $$1800 \text{ lb/in}^2 \times 14.2 \text{ in}^2$$
   $$= 25,560 \text{ lb f}$$
5. Calculate force on annulus side of piston.
   Force on annulus side = annulus pressure × annulus area

   $$1800 \text{ lb/in}^2 \times 8.25 \text{ in}^2$$
   $$= 14,850 \text{ lbf}$$
6. Force on shot plunger = 25,560 - 14,850 = 10,710 lb.
7. Calculate pressure on the metal from the shot plunger diameter to give the following results.

Shot plunger diameter, in	Pressure on metal, lb/in^2
1½	6061
2	3409
2½	2181
3	1515

The above results were obtained on a machine with the intensifier in the "off" position, and the machine was of a design which maintained a balance pressure on both sides of the hydraulic injection piston during injection. The force on the injection ram is seen to be due to the different areas of the head and annulus sides of the piston.

[Other machines may not operate in this way, and are based on pressure differences on the head and annulus sides of the hydraulic piston but the same method of calculating the force on the shot plunger is used.]

If the intensifier had been in the "on" position, the force would have been increased. With an intensification ratio of 1:1.5, the head side pressure would have been 2700 lb/in^2, if 1:2 it would have been 3600 lb/in^2. These increases would have given an increase in force on the injection ram and a greater pressure on the metal after cavity-fill.

The ability to vary the final pressure on metal by bringing in intensification and by changing the shot sleeve diameter can give a wide range of final pressure variations. It should be remembered however that the pressure calculated from the force on the shot plunger is the nominal value, and not necessarily the true metal pressure during the cavity fill period. The two are the same only if there are no pressure losses due to metal turbulence, changes in metal direction or changes in the dimensions of the injection system and this will be discussed later. Also if the runner system has solidified before the casting, no pressure will be transmitted to the casting at the end of the injection period. It is most important to bring the intensification in at exactly the correct time in the casting cycle; if too late, the extra pressure will be ineffective, and if too early, the die may "flash". The use of instruments to check the actions of the intensifier is, therefore, a very great asset. When real traces are being measured, there are many instances of uncertainty as to the actual pressure to be measured. Fluctuations are seen to occur most commonly during the cavity-fill period or immediately after cavity-fill due to inertia effects. Only experience will indicate whether such effects should be ignored or considered in the calculations.

Displacement Measurements

There are three useful diecasting parameters which are calculated from the displacement transducer trace. These are as follows:

1. Plunger velocity during first and second stages of injection, or under dry shot conditions.
2. Velocity of metal through the gate into the die cavity, i.e. gate velocity.

3. Time to fill the die cavity, i.e. cavity-fill time and overflow system.

Plunger Velocity

This is usually very easy to measure, the methods having previously been described. Plunger velocity is proportional to the angle of the trace, which is measured; the plunger velocity is read off from the master graph for the paper speed being used. Figure 3.20 is an idealised situation, but with actual traces there are differences; only with experience in interpreting traces will their cause be established. Two more common variations are due to "plunger creep" and to a slowing down of the second stage injection speed during injection.

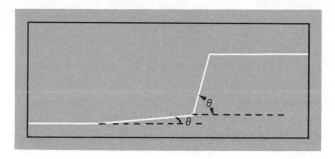

FIG. 3.20 VELOCITY OF 1st AND 2nd STAGES IS PROPORTIONAL TO THE RELEVANT ANGLES

"Plunger creep" only occurs on hot chamber machines and it is due to molten zinc leaking past the plunger and piston rings and allows the plunger to keep moving after the die cavity is full. Figure 3.21 is a displacement trace with "plunger creep", where the continued plunger movement is shown as a further sloping line after the end of cavity-fill. A certain amount of "plunger creep" is common on hot chamber machines (in practice the alternative may be to slow down or seize up the plunger due to tight clearances), and it has little or no effect. However, serious "plunger creep" can lead to the injection piston "bottoming out", and under these conditions no pressure would be applied to the die cavity and defective castings would probably result.

Slowing down during the second stage is most common if the first to second stage speed changeover point occurs well before the metal reaches the gates at the casting. The initially high second stage speed is slowed down by the gate restriction. In these instances the speed to be measured is the speed during cavity-fill, not the initially faster second stage speed.

Gate Velocity

This has a simple arithmetical relationship to second stage plunger velocity as follows:

$$\text{Gate velocity} = \frac{\text{Plunger area}}{\text{Gate area}} \times \text{Plunger velocity}$$

The plunger area is known because the size of the shot cylinder in which it moves is known. The plunger velocity is obtained as described above, and the gate area is obtained from either the die drawing or by accurate measurement on an actual casting/runner assembly.

For example, for the injection system shown in Fig. 3.8, the plunger areas = 3.1 in². Let us assume that the plunger velocity as measured is 8 ft/sec and the total gate area is 0.15 in².

Plunger area = 0.0215 ft²
Gate area = 0.00104 ft²
Plunger velocity = 8 ft/sec

$$\text{Therefore Gate velocity} = \frac{0.0215}{0.00104} \times 8 = 165 \text{ ft/sec}$$

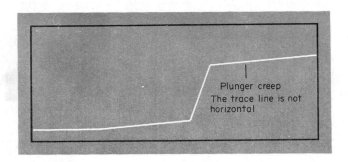

Plunger creep
The trace line is not horizontal

FIG. 3.21 DISPLACEMENT TRACE SHOWING PLUNGER CREEP

Die Cavity-Fill Time

This is the time taken to fill the die cavity and overflow wells attached to the casting. To obtain it we first have to obtain the volume of the casting and overflow system and then calculate the length of shot plunger travel which corresponds to this volume. This distance is then marked on the trace vertically downwards from the end of the cavity-fill period. A horizontal line is then drawn from the end of this distance until it intersects the displacement trace. The time as measured on the trace, between this point of intersection and the end of cavity-fill, is the cavity-fill time.

For example, To obtain volume from casting weight we divide the weight by the specific gravity of the liquid metal which for zinc = 6.5 g/cm^3 and aluminium = 2.5 g/cm^3.

For an aluminium casting of weight 725 g the volume

$$= \frac{725}{2.5} = 290 \ cm^3$$

To obtain the plunger stroke corresponding to this volume, we divide by the plunger area. For the system in Fig. 3.8, the plunger area = 20 cm^2.

$$\text{Therefore Plunger stroke} = \frac{290}{20} = 14.50 \ \text{cm in length}$$

Mark on the trace, the distance corresponding to 14.50 cm (Fig. 3.22) downwards from the end of the cavity-fill period.

Draw a horizontal line from this point to intersect the displacement trace line.

Measure the time from this point of intersection to that at the end of cavity-fill. This is the cavity-fill time.

Measurements of the above usually show that the changeover from first to second stage injection occurs before the metal is at the gates to the casting cavity, and this is to be expected. If the changeover occurred after the metal had reached the gates, it would mean that metal was being injected into the casting cavity at a very slow velocity initially. This velocity would be increased as the second stage came into opertion but defective castings would probably result because the earlier injected metal may have solidified and not be remelted by

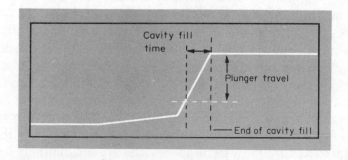

FIG. 3.22 MEASUREMENT OF CAVITY-FILL TIME

the subsequent metal. Figure 3.23 shows the type of trace obtained and defective castings may result. Since the changeover between first and second stage of injection is often controlled by a movable trip switch, the use of instruments and the above measurements can be of great value in its correct placement.

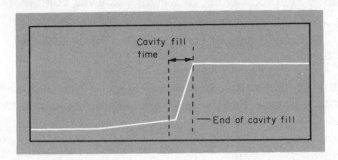

FIG. 3.23 SHOWING 1st AND 2nd STAGE CHANGEOVER OCCURRING DURING THE CAVITY-FILL PERIOD

THE CHECKING OF DIECASTING MACHINES

The performance of the injection system of a machine is measured for three basic reasons, these are as follows:

1. To check the performance of a new machine.
2. To continually check a machine to see if performance is variable and to assist in die setting and quality control.
3. For trouble shooting puproses.

Checking of a New Machine

It is advisble to check a new machine in order to ensure that the machine is to the maker's specification with regard to pressures, injection velocity, etc.

Checking should be systematically carried out both with a die in place, making castings and also on "dry shots". "Dry shots" measure the performance of the injection system in the absence of molton metal, and for cold chamber machines they are made by inserting a ball of cloth in the shot sleeve and firing this against the die block with the injection plunger. For hot chamber machines, the plunger in the gooseneck should be removed and a strong baulk of wood placed across the gooseneck aperture. The injection ram can then be fired against the baulk of wood. It is worth noting that with no metal as a restraint, injection velocities are substantially increased for the same valve settings so it is advisable when checking to start at the slower velocity settings. With a new machine all the various control wheels or knobs which effect the various phases of injection with regard to velocity and pressure should be checked to establish the response they produce on the U.V. oscillograph trace. Since it is not unknown for the setting of one valve to effect the response of another, even though the latter valve has not been touched, it is essential to check for this type of interplay.

The following checking calibration routine is suggested.

1. 1st stage plunger velocity versus 1st stage setting.
2. 2nd stage plunger velocity versus 2nd stage setting.
3. The setting on the valve controlling intensifier build-up time versus the measured build-up time at maximum and minimum 2nd stage injection speeds.
4. The setting on the valve delaying start of intensification versus measured delay time.
5. Any other valve or combination of the above which will influence the above parameters of the injection system.
6. Measure pressures and calculate the forces on the injection piston.

When carrying out the above calibrations, the hydraulic fluid type and temperature should be noted, as also should the injection line pressure and accumulator precharge pressure. It may also be necessary to repeat 1 - 6 using various pressures.

The type of results obtained may be plotted in a similar way to Figs. 3.24-3.26.

Figure 3.24 shows the first-stage velocity using a rag shot for two nominally identical machines. It will be seen that they have a far from identical response to the same setting of the control valve.

Figure 3.25 shows the difference in plunger velocity when using rag shots and when casting (*N.B.* Different dies will produce different effects as is shown.)

Figure 3.26 shows the interrelationship between control valves in different parts of the hydraulic circuit. Here it is seen that the time taken to reach the required intensification pressure is influenced by the second stage injection speed.

The checking of a machine as described is generally undertaken with a die in place, while castings are being made and some of the results obtained can be influenced by the particular die on the machine as well as by the machine performance itself. It is necessary, therefore, in addition to the above checks, to be able to measure the performance of a machine in isolation from the effects of the die. This is done by constructing a machine performance diagram,

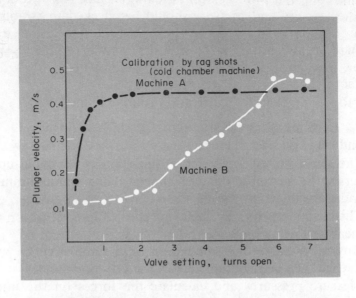

FIG. 3.24 PHASE 1 PLUNGER VELOCITY VALVE SETTING VS PLUNGER VELOCITY

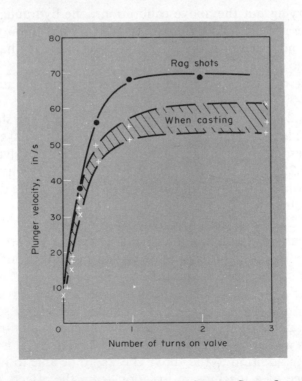

FIG. 3.25 EFFECT OF SPEED CONTROL VALVE ON SECOND STAGE INJECTION
VELOCITY

sometimes called a PQ^2 diagram, whereby the pressure developed in the injection system (P) is related to the quantity (Q) of metal delivered. The term Q^2 is, however, plotted in preference to Q, since the relationship between P and Q^2 is a straight line for hydraulic

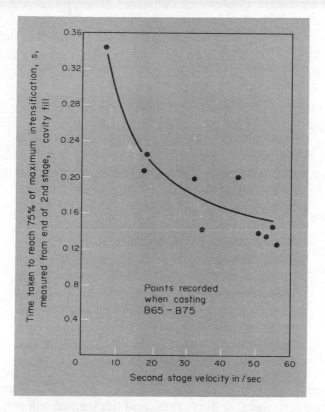

Fig. 3.26 Relationship Between Second Stage Velocity and Rate of Intensification

machines which makes using the diagrams easier. In practice, the diagrams are generally shown with Q plotted on a square scale rather the Q^2 plotted on a linear scale. PQ^2 diagrams are commonly presented in one of three forms as follows:

1. Metal pressure vs. Metal delivery rate (l/sec).
2. Metal pressure vs. Second stage injection velocity.
3. Machine accumulator pressure or air pressure for pneumatic machines vs. Second stage injection velocity.

The graph of metal pressure vs. metal delivery rate (1/sec) is the most useful form and is the graph used by die designers when they are matching the characteristics of a machine to the requirements of the die they are designing. However, all three forms of the graph are arithmetically related to each other by the forces on the injection piston, shot plunger and shot plunger diameter and having obtained one graph the others can be calculated.

The information required to construct graph number 3 should be available from the machine makers, since they should know the two parameters involved. It may be that the data given is for the machine type rather than for the actual machine being purchased and in these cases a check on the actual machine should be carried out.

Construction of a Diecasting Machine Injection Performance Diagram, i.e. a PQ^2 Diagram

This is a reasonably straightforward exercise for a technician to undertake, using the same methods and instruments as previously described. The necessary steps are as follows:

1. If an intensifier is fitted, turn to the "off" position. Measure the pressure at the head and annulus side of the injection piston at the end of the cavity-fill period, i.e. the plunger is at rest.
2. Calculate the forces on the injection piston as previously described.
3. Calculate the pressure on the metal by dividing the force on the injection piston by the shot plunger area.
4. Measure the maximum velocity of the plunger under "dry shot" conditions.
5. This will provide two datum points. The first will be the maximum pressure on the metal at zero plunger movement, and hence zero metal delivery rate, and the second will provide maximum plunger velocity (and hence maximum delivery rate) at zero metal pressure, since under "dry shot" conditions there is no metal in the system.
6. A graph should be constructed, the vertical axis being metal pressure, the horizontal axis being plunger velocity^2 or metal delivery rate^2 and the above two points marked on the appropriate axis.
7. For hydraulic machines a straight line should be drawn between the two points. For pneumatic machines it is a curve, and further points will be needed before the curve can be drawn; the method of obtaining these will be demonstrated later. This is the injection performance diagram or PQ^2 diagram for the machine.

8. Figure 3.27 shows a typical diagram and the line represents the limits of the machine performance relative to metal pressure and plunger velocity[2]. Castings which require velocities and pressures at values below the line can be made on the machine, whilst those requiring velocities and pressures above the line cannot be made on the machine.

The castings are not necessarily always made at the maximum performance of the machine, and the variations in machine performance below maximum can be obtained by the controls on the machine relative to pressures and injection velocity.

Figure 3.28 shows the effect of varying the shot plunger diameter. This will have no effect on the dry shot injection velocity, but it will affect metal pressure. However, if the graph is plotted in a different way, i.e. metal delivery rate[2] instead of plunger velocity[2], the effect of shot plunger diameter is more easily recognised (Fig. 3.29).

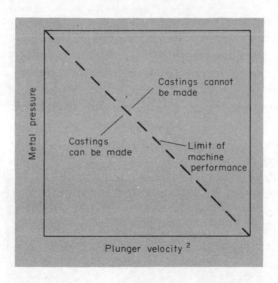

FIG. 3.27 TYPICAL PERFORMANCE DIAGRAM

The construction of an actual diagram is described as follows, and since the machine was pneumatic, the method of obtaining additional points on the machine characteristics curve is also indicated.

1. The injection system of the machine is as shown in Fig. 3.30.
2. The maximum pressure on the metal was 1200 lb/in[2].
3. The maximum plunger velocity (dry shot) developed was 3.1 m/sec.

4. If this had been a hydraulic machine, the performance graph would have been a straight line joining 1200 lb/in^2 at zero plunger velocity to 3.1 m/sec plunger velocity at zero metal pressure.

5. Since the machine is pneumatic, we need extra points to obtain the performance curve and one such point is obtained by fitting a die and making a casting at the maximum plunger velocity

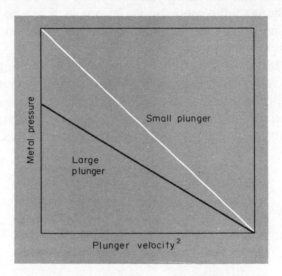

FIG. 3.28 EFFECT OF CHANGING PLUNGER DIAMETER

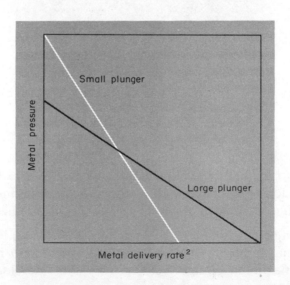

FIG. 3.29 EFFECT OF PLUNGER DIAMETER ON METAL DELIVERY RATE

possible, and measuring the metal pressure during the cavity-fill period. In this case the plunger velocity was 1.1 m/sec and the metal pressure 850 lb/in². Thus a further point on the curve was obtained at the intersection of 850 lb/in² and 1.1 m/sec and the curve was drawn (Fig. 3.31) to obtain further points

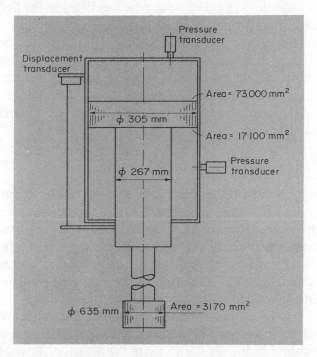

FIG. 3.30 ARRANGEMENT OF INJECTION CYLINDERS, PLUNGER AND TRANSDUCER POSITIONS ON MACHINE

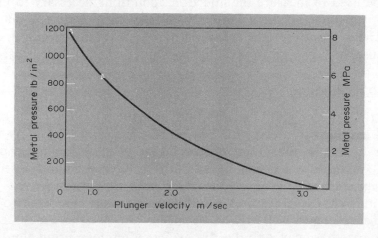

FIG. 3.31 MACHINE INJECTION CHARACTERISTICS

on the curve, different dies would have to be fitted and the maximum plunger velocities and metal pressures during cavity-fill measured and plotted on the graph. In practice these further points can be measured when different dies are put on the machine during normal production and if necessary the original curve can be modified in the light of these further results.

It should be noted that in Fig 3.31, plunger velocity has been plotted on a squared scale rather than plunger velocity^2 on a normal scale. In practical die design the PQ^2 diagram is more useful if instead of plunger velocity, metal delivery rate is plotted. The two are related by the shot plunger diameter (Table 3.2).

Routine Checking of Machines and Die Setting

Having established the facts on a new machine, the various parameters can be checked at intervals to see if there is any change. Valves can wear, velocities become reduced and pressures can fall, etc., and such phenomena may influence casting quality. Instruments can have advantages if incorporated into a recognised inspection and quality control system. The setting of dies can also be achieved using instruments. Every foundry has its share of "difficult" dies which are sensitive to injection velocities, pressures, die temperatures, etc., and setting such dies can be a time-consuming trial-and-error exercise, particularly if the die has to be mounted on different machines. The use of instruments can eliminate this trial-and-error. Instruments can be used to measure the injection conditions necessary to make good castings, and these can be reset each time the die is put up. Similarly, the same conditions with regard to injection velocity, pressure, etc., can be set on other machines when the die is used.

Trouble Shooting

If time is available, the injection conditions necessary to make good castings from every die can be obtained and filed away with other data on the die. Often this is achieved by taking a photocopy of an U.V. trace which can be used for reference should defective castings start to be produced in unacceptable numbers.

In this case several traces would be taken and compared to this original trace. Any differences in speeds and pressures may be obvious, or they may only be discovered after measuring the trace parameters. If differences are discovered, the machine can readily be

TABLE 3.2 RELATIONSHIP BETWEEN PLUNGER VELOCITY, PLUNGER DIAMETER AND METAL DELIVERY RATE

Metal delivery rate l/sec

Plunger Velocity m/sec	Shot plunger dia mm											
	30	40	50	60	70	80	90	100	110	120	130	140
0.5	0.35	0.63	0.98	1.42	1.93	2.52	3.18	3.93	4.75	5.66	6.64	7.7
1.0	0.71	1.26	1.96	2.83	3.85	5.03	6.36	7.86	9.5	11.31	13.28	15.39
2.0	1.41	2.52	3.96	5.66	7.7	10.06	12.72	15.72	19	22.6	26.56	30.78
3.0	2.12	3.78	5.88	8.49	11.55	15.09	19.08	23.58	28.5	33.93	39.84	46.17
4.0	2.83	5.04	7.84	11.32	15.4	20.12	25.44	31.44	38	45.24	53.12	61.56
5.0	3.54	6.3	9.8	14.15	19.25	25.15	31.8	39.3	47.5	56.55	66.4	76.95
6.0	4.24	7.56	11.56	16.98	23.1	30.18	38.16	47.16	57	67.86	79.68	92.34
7.0	4.95	8.82	13.72	19.81	26.95	35.21	44.52	55.02	66.5	79.17	92.96	107.7
8.0	5.66	10.08	15.68	22.64	30.0	40.24	50.88	62.88	76	90.48	106.2	123.1
9.0	6.36	11.34	17.64	25.47	34.65	45.27	57.24	70.74	85.5	101.8	119.5	138.5
10.0	7.07	12.6	19.6	28.3	38.5	50.3	63.6	78.6	95	113.1	132.8	153.9

reset to the original condition, that is, unless it has developed a fault. If no differences in the traces can be detected even after measurement, it is possible that the trouble may rest with the die in regard to variables such as temperature, blockage of vents, etc. In any event the use of instruments will separate variables due to the machine and those due the die and enable personnel to concentrate attention in the particular area where it is needed, rather than the frequently employed trial-and-error approach whereby every available control is adjusted in an effort to produce good castings.

CONSTRAINTS TO INJECTION SYSTEM EFFICIENCY

Students of hydraulics may be tempted to ask why it is necessary to measure both metal pressure and velocity, when the two are related by the Bernouilli equation.

$$V = Cd\sqrt{\tfrac{2}{\rho}p}$$

where V = actual metal velocity, m/sec,
  Cd = discharge coefficient,
  ρ = molten metal density, kg/m^3,
  p = metal pressure (pa).

The reason is that the discharge coefficient is not necessarily known. The discharge coefficient is a measure of the efficiency of the system with regard to metal flow and is influenced by the surface condition of the passages in the gooseneck, turbulence and the number and extent of the changes in direction which the metal has to take from the shot plunger to the die, etc. The discharge coefficient is the ratio

Metal velocity achieved at the gate
_____

Theoretical metal velocity at the gate [i.e. with no pressure losses]

and this ratio is always numerically less than 1.0.

For hot chamber machines a typical Cd value is ~0.65, whilst 0.85 is relatively high and 0.45 is low.

For the commonly cast alloys of zinc, aluminium, magnesium and brass, Bernouilli's equation can be written as follows:

For zinc alloys $Cd = \dfrac{55.4V}{\sqrt{p}}$

For aluminium alloys $Cd = \dfrac{35.4V}{\sqrt{p}}$

For magnesium alloys $Cd = \dfrac{28.5\,V}{\sqrt{p}}$

For brass $\qquad\qquad Cd = \dfrac{61.6\,V}{\sqrt{p}}$

To calculate the Cd value, the actual metal pressure and velocity at the gate have to be measured, and this is carried out as previously described. The values are then inserted in the above equations. If we take as an example a typical gate velocity of 40 m/sec and a metal pressure of 8,000,000 pa (i.e. 8 Mpa) for a zinc casting we have the following situation:

$$Cd = \frac{55.4\,V}{\sqrt{p}}$$

$$Cd = \frac{55.4 \times 40}{\sqrt{8,000,000}}$$

$$Cd = 0.78.$$

If we had no pressure losses, i.e. $Cd = 1$, the pressure required to give a gate velocity of 40m/sec is as follows:

$$Cd = \frac{55.4\,V}{\sqrt{p}}$$

$$p = (55.4V)^2$$

$$p = 4,910,656 \text{ pa}$$

$$p = 4.9 \text{ Mpa.}$$

The difference between 8 Mpa for a Cd of 0.78 and 4.9 Mpa for a Cd of 1.0, i.e. ~3 Mpa, is due to the pressure losses in the injection system.

Taking the extreme values previously mentioned, i.e. Cd's of 0.45 and 0.85, it can be calculated that the metal pressure needed to give a gate velocity of 40 m/sec are 24.2 Mpa and 6.8 Mpa respectively. It has been calculated[2] that improving the Cd from 0.45 to 0.85 would reduce the cavity-fill time by 20%.

It can be seen, therefore, that attention by the machine maker to gooseneck and nozzle proportions and design can be a worthwhile exercise and will benefit the diecaster, since a higher Cd value would indicate that the metal pressure can be reduced for the same injection velocity or alternatively a faster injection velocity can be obtained for the same pressure.

There is, however, the law of diminishing returns. To improve the Cd from 0.45 to 0.85 would reduce the cavity-fill time by 20% whilst from 0.65 to 0.85 an improvement of only 7% would result.

As compared to the hot chamber process, where metal has several changes of direction in the gooseneck and nozzle, the cold chamber process has a much more direct injection system. This should give much higher Cd values, which in conjunction with the lower density of aluminium would indicate that lower metal pressures should be needed.

The factors influencing the Cd value of cold chamber systems appears to have been much less well documented than for hot chamber systems, but what has been recorded would indicate that the Cd values can be quite low.[2] The Cd value is seen to be very dependent on the time delay between filling the shot sleeve and pressing the injection button and upon the state of wear of the shot plunger. Presumably boundary lubrication in the shot plunger and shot end alignment must also be of importance.

It has been shown that a delay of 2 sec after filling the shot sleeve before injecting the metal can reduce the metal velocity by 13%, whilst a delay of 4 sec can reduce it by 46%.

If we use the Bernouilli equation for aluminium

$$Cd = \frac{35.4V}{\sqrt{p}}$$

and make the assumption that the $Cd = 0.9$ for zero delay between pouring and injection for a gate velocity of 40 m/sec, it can be calculated that the metal pressure should be ~2.4 Mpa.

If, however, the injection velocity is reduced by 46% due to a delay in injecting the metal, the gate velocity is lowered to 21.6 m/sec. and if the pressure is retained at ~2.4 Mpa, it can be calculated that the Cd value is 0.49. Alternatively, to maintain the same gate velocity the metal pressure would need to be ~8.3 Mpa.

The reason for the low Cd with delayed metal injection is probably due to partial freezing in the shot sleeve, offering a restriction to metal flow. With a worn plunger, metal pressures of ~50% higher have been found to be needed for the same gate velocity. In this case the increased pressure is needed to overcome the friction caused by flash or metal skull jamming at the plunger/shot sleeve interface.

Both the question of delayed injection and worn plungers are factors under the control of the foundry concerned, rather than the machine maker. Plunger wear can, however, be much reduced by very accurate alignment of the shot end during machine manufacture, which is best achieved by optical methods. If this is combined with the facility to realign the shot end, should this be necessary on the foundry floor, a much better machine will result.

Locking System

FUNCTION AND ACTION OF THE LOCKING MECHANISM

The locking mechanism places a preload on the die such that the force across the die-half interfaces is greater than the force due to the metal pressure at injection. If there were no force across the die, or if the force were insufficient, the metal injection forces would cause the die to "flash", i.e. metal would be ejected at the die-half interface. In practice, the locking force has to be sufficiently great to balance out the metal pressure, pressure due to inertia effects in the injection system (similar to "water hammer" in pipes), as well as higher pressures exerted in certain areas of the die due to an offset die cavity.

The geometry of the toggle system is such that there is not a constant ratio of locking cylinder piston movement to platen movement, rather the opposite. As the die closes there is an increasing ratio of piston movement to platen movement so that the mechanical advantage gained increases the nearer the dies are to the locking over point.

MEASUREMENT OF LOCKING CYLINDER PRESSURE FORCE

The movement of the locking system during the locking action is much slower than the injection system when making a casting and consequently it is often considered not so important to have a fast response method of measuring pressure. Generally speaking, machines have only a pressure gauge fitted which either monitors continuously the changes in pressure during the cycle or is actuated by a push button as required. This method is imprecise because the pressure gauge will only accurately measure the final line pressure after the locking has been completed, and this is not the most important pressure to measure.

Figure 3.32 shows the pressure changes in the locking cylinder during locking. A characteristic peak and relaxation in pressure is seen to occur, which represents the initial closure and compression of the die with the relaxation of pressure, reflecting the increased mechanical advantage of the toggle linkage system. When the system is locked, the pressure is seen to build up again to the maximum pressure available, which is the pressure normally measured by any gauge fitted by the machine maker.

To achieve the maximum locking force, the initial peak pressure and the final pressure must be the same, i.e. maximum pressure must occur at the point of maximum mechanical advantage. If the lock is reduced by adjustment of the machine, the initial peak pressure is lower since the force required is less (Fig. 3.33). Generally speaking, a machine is set to its maximum locking force, and a check on the peak and final pressures will show whether this has been achieved. If the two pressures do not coincide, the machine is operating below its maximum lock for the line pressure being used. It should be remembered that even though the peak and line pressures coincide it does not necessarily mean that the locking force is at the stated value for the machine, it simply indicates that it is the maximum attainable at that particular line pressure and condition of machine.

The pressure trough between the initial peak and final pressure is also of interest, since its value may be an indication of the friction in the system.

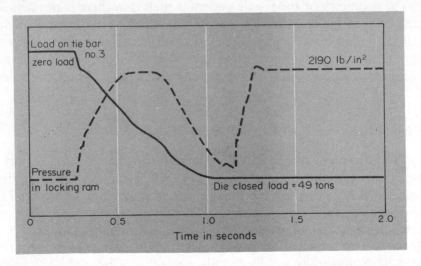

FIG. 3.32 RECORDING OF THE PRESSURE INSIDE THE LOCKING RAM AND THE LOAD ON ONE OF THE TIE BARS (PROPORTIONAL TO LOCKING FORCE) WHEN THE MACHINE WAS SET AT THE MAXIMUM LOCKING FORCE FOR THE GIVEN LINE PRESSURE

MEASUREMENT OF LOCKING FORCE

The locking force of a machine is fundamental to performance and is commonly the criterion by which it is purchased. In the United Kingdom most machines are in the range ~50 - 800 tons locking force, but there are machines available of from 5 to 3000 tons. Of equal importance to the total locking force is its even distribution over the tie bars. Uneven tie bar loading can cause high stresses elsewhere in the system and lead to toggle damage and the stretching or even fracture of tie bars. It has been said[6] that toggle wear can be increased by

50% for a 25% overload. Unfortunately overload can cause increased wear and wear can cause increased overload, and once a machine is seriously out of adjustment the above sequences can probably cause rapid deterioration. The measurement of locking force and its distribution over individual tie bars has not received great attention, but where it has been investigated it has shown cause for concern.[6, 7]

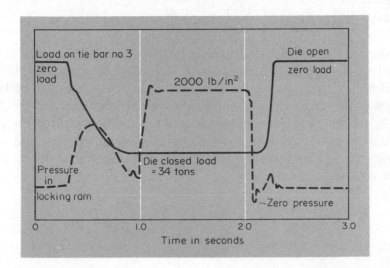

FIG. 3.33 RECORDING OF THE PRESSURE INSIDE THE LOCKING RAM AND THE LOAD ON ONE OF THE TIE BARS (PROPORTIONAL TO LOCKING FORCE) WHEN THE MACHINE WAS SET BELOW MAXIMUM LOCK FOR THE GIVEN LINE PRESSURE

Values of >20% of proportional load have been quoted due to unevenly worn linkages, non-parallel dies, off-centre mounted dies, etc. Another main reason for maladjustment is loss of tie bar synchronisation when one bar is completely removed for fixing large dies. It has been said[6] that deviations up to 80% from correct loading have occurred and remained undetected until actual measurements were made. One machine which was thoroughly investigated[7] gave the results detailed in Table 3.3, where it will be seen that out of a total locking force of 105.6 tons the load was distributed over the tie bars as follows:

Front upper 45.4%
Front lower 17.7%
Rear upper 10.8%
Rear lower 26.1%

Table 3.4 illustrates that when the load is out of balance equal movements of the tie bar adjusting nuts do not have an equal effect. Eventually a balanced load in the tie bars was achieved by having zero nut settings as below as compared to the initial zero nut settings.

> Front upper - 1½ divisions
> Front lower + 4 divisions
> Rear upper + 14½ divisions
> Rear lower + 11½ divisions

In addition to a tensile stress in each tie bar, there is frequently a bending stress. Bending stresses arise because the machine platens deform elastically during the locking operation. The extent of bending is governed by the platen rigidity and the load imposed on each bar by the physical size and location of the die between the platens. Bending does not occur in any particular plane in a tie bar and only actual strain gauge measurements taken radially around the bar will show what is happening.

TABLE 3.3 MEASURED LOADS ON TIE BARS

Tie bars	Load, tons	Mean tensile stress, tons/sq'' in^2
Front upper	47.9	3.0
Front lower	18.7	1.24
Rear upper	11.4	0.72
Rear lower	27.6	1.73

TABLE 3.4 EFFECT OF ADJUSTMENTS ON OUT-OF-BALANCE TIE BARS

Tie Bar	Load (tons) after adjustment	Increase	Divisions on nut adjustments	Increase/ division on nut
Front upper	58.4	10.5	+ 2	5.3
Front lower	43.8	25.1	+ 5	5.0
Rear upper	30.4	19.0	+ 5	3.8
Rear lower	39.8	12.2	+ 5	2.4

Methods of Measurement

There are two basic ways to measure locking-force which are:

1. To measure the reaction in the individual tie bars, the sum of which is the total locking force.
2. To measure the reaction between the platens using a load cell.

Measurement of Tie Bar Strain

Strain can be measured in individual tie bars by either the use of strain gauges or extensometers, which, with a knowledge of Young's Modulus of Elasticity for the tie bar, will enable the stress to be calculated.

The advantages of this method are:

1. An approximation of the locking force distribution over the die can be obtained.
2. The locking force can be measured with dies of different sizes and tie bars can be individually adjusted to give the same stress in each bar.
3. The locking force can be measured during casting.

A disadvantage of the strain gauge method is bonding the strain gauges to the tie bars and the associated electrical wiring and equipment which is usually of a temporary nature.

Cost is relatively high; each bar will need four strain gauges mounted and wired such that tie bar bending moment is compensated.

To overcome these disadvantages, strain gauges can be mounted onto steel thrust rings which fit under each tie bar nut. Four are required for each machine and some diecasting machines are supplied with this system as an optional extra. A four-way switch is usually provided to enable the four readings to be measured on one instrument. The disadvantage is that the rings are not readily transferable between machines.

Extensometers measure the increase in actual length of the tie bar during locking and are based on either micrometer or dial gauge readings taken over the complete length or a selected portion of the tie bar. If over the latter, usually some form of lever device is incorporated to magnify the actual reading. The order of magnitude of tie bar change in length will depend upon the tie bar diameter, length and general rigidity of the machine, but a typical change would be ~ 0.030 in.

The inside micrometer method measures the distance between the fixed and thrust platen at each corner, the actual measurement being read on a dial gauge or micrometer head.

Another method using a dial gauge is to measure the movement on the tie bar ends. Each tie bar has a hole drilled centrally from one end into the portion of the bar which is stressed under lock. A dial indicator is mounted at the end so that its spindle rests at the bottom of the hole. During locking the tie bar is stretched whilst the spindle is not, and the difference in readings is a function of the extension. A dial gauge is also the basis of a strap-on or magnetically attached device which measures the extension over a short length of tie bar. The major disadvantage of this type of measurement is because of the bending of the tie bar, and several measurements should be taken at various radial positions on the tie bar if the instrument is used between the fixed and moving platen. Bending is, however, less between the thrust and moving platen, and measurements here should be more independent of radial position. The difficulty here is the practicability of access due to guards on the machine.

Measurement of Platen Reaction

Platen reaction is usually measured with a load cell mounted between the platens as though it was a die block (Fig. 3.34). The load cell is typically in the form of a steel cylinder on which strain gauges are mounted on the inner walls. Locking force is measured by resistance changes resulting from compressive strain in the steel cylinder. Several cylinders are needed to cater for the various sizes of machines and standard sizes have been established in the United States and the United Kingdom (Tables 3.5, 3.6). Different sizes are needed to keep tie bar bending stresses to a minimum, since these would be high if a machine having a large platen size was tested with a small diameter cell.

The advantage of the load cell method of testing is that it exactly simulates a real die, and also the speed at which a machine can be tested and returned to production. The disadvantages are that only the total locking force is measured and not the distribution on the individual tie bars, and also that the die has to be removed prior to fixing the load cell.

RELATIONSHIP BETWEEN LOCKING PRESSURE AND LOCKING FORCE

Since machines are usually specified by their locking force, the pressure to achieve the locking force must also be specified. The rela-

FIG. 3.34 LOAD CELL MOUNTED BETWEEN PLATENS

TABLE 3.5 STANDARD LOAD CELLS USED IN U.K.

Locking force range (tons)	Load cell dimensions (in)		
	O.D.	I.D.	L.
30 - 70	9	8.2	9
70 - 160	11	9.6	11
160 - 375	14.5	11.9	15
375 - 870	20	15.5	21.5

TABLE 3.6 LOAD CELLS RECOMMENDED BY S.D.C.E.

Locking force	Load cell dimensions (in)		
range (tons)	O.D.	I.D.	L.
35 - 150	8	5	8 - 10
75 - 350	13	10	8 - 12
200 - 500	18	14½	12 - 18
350 - 1000	28	24	20 - 30
750 - 2000	39½	34	34 - 50

tionship between pressure and locking force for a 200-ton machine is shown in Fig. 3.35 where it is seen that the 200 tons locking force is achieved at ~2100 lb/in^2, graphs such as this should be obtained for each machine purchased. If we consider what happens to a machine when it is being locked, i.e. the machine is being stretched, it is seen that the relationship between pressure and locking force is not necessarily linear. The types of relationship possible are shown diagramatically in Fig. 3.36 where a typical modern machine is compared to both a "hard" and "soft" machine. With a "soft" machine the locking force reaches a value above which it will not increase, despite increasing the locking pressure, since the machine is being stretched. This is an undesirable feature, since the curve may level of at values not substantially above the rated locking force of the machine. However, a machine such as this will not be sensitive to variations in locking pressure which can occur when wear develops in the hydraulic pump, etc.

A "hard" machine, on the other hand, will readily achieve the necessary locking force, but will be sensitive to changes in pressure. A modern machine is, therefore, a compromise between the two extremes, but it is important that the relationship is known.

RELATIONSHIP BETWEEN LOCKING FORCE AND TIE
BAR ADJUSTMENT

This is a most important parameter in practical diecasting, since a machine which is overset will not lock, whilst a machine which is underset may possibly flash the die. If a machine is set in the cold condition, when they warm up, the machine and die will expand to operating temperature and the lock will have to be reduced. It is im-

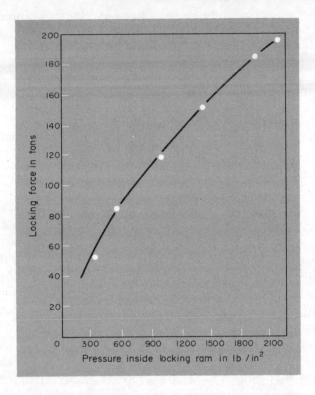

FIG. 3.35 EFFECT OF LINE PRESSURE ON THE LOCKING FORCE OF A 200
TON MACHINE

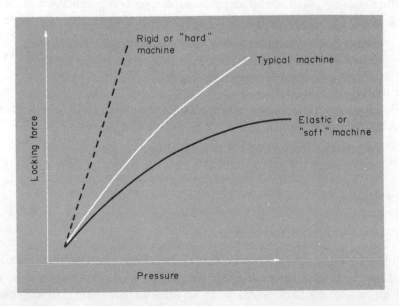

FIG. 3.36 RELATIONSHIP BETWEEN LOCKING FORCE AND LOCKING
CYLINDER PRESSURE

portant to know the relationship between locking force and the graduations on the tie bar adjustment nuts. Figure 3.37 shows diagramatically the type of relationship to be expected. A "soft" machine will be less sensitive to the tie bar nut settings, whilst a "hard" machine will be more sensitive, and lock will quickly be lowered as the tie bar nuts are released. As before, the best type of machine is a compromise between the two extremes.

Many modern machines have a centralised tie bar adjusting system operated by an electric motor, in which case the above relationship is difficult to obtain.

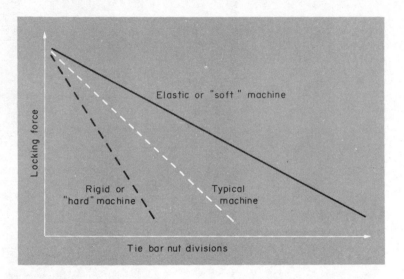

FIG. 3.37 RELATIONSHIP BETWEEN LOCKING FORCE AND THE TIE BAR NUT ADJUSTMENT

MATCHING A DIE TO THE LOCKING FORCE

There are three factors to be considered for which allowances should be made.

1. Allowance for an offset die cavity.
2. Allowance for the machine being set below maximum lock.
3. Allowance for pressure peaks in the injection system during casting.

Allowance for Offset Die Cavity

Ideally a die should fit centrally between the tie bars, and the casting spray should also be central. This is not always possible, and an allowance has to be made to overcome the effect of off-centre loading. Because of such features as machine rigidity and actual die size, it is not possible to calculate accurately what is involved; however, an approximation can be calculated as follows:[2]

1. Calculate the position of the centroid of the casting spray which is done by taking moments of area as shown in Fig. 3.38.
2. Use the following equation to find the required locking force:

Required locking force $F = P \left(1 + \dfrac{2e}{100}\right)$

where $P =$ the die parting force produced by the metal pressure in the die cavity,

$e =$ the largest percentage offset (either horizontal or vertical) of the centroid of the impression area.

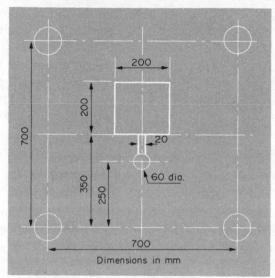

Dimensions in mm

A Area mm²	B Distance of centroid from centre line bottom tie bars mm	C = A × B Moment of area about Centre line bottom tie bars mm³
Slug 2827	250	706,750
Runner 1400	315	441,000
Casting 40000	450	18,000,000
Total 44227		19,147,750

Centroid of spray area from centre line bottom tie bars

$= \dfrac{C}{A} = \dfrac{19,147,750}{44227} = 432.9$ mm

Vertical off-set $= 432.9 - \dfrac{700}{2} = 82.9$ mm

% vertical off-set $= \dfrac{82.9}{700} \times 100 = 11.8$

% horizontal off-set $= 0$.

FIG. 3.38 METHOD OF CALCULATING PROJECTED AREA OFF-SET.

For example shown in Fig. 3.38 where the largest offset is 11.8%

$$F = 1.24p$$

i.e. a 24% greater locking force is required than if the casting spray were in the centre of the die.

Allowance for Machine Set Below Maximum Lock

It has been seen previously how the locking force can be influenced by variations in pressure and tie bar nut adjustment, and it is at this stage in the calculations that graphs are required of locking pressure and tie bar nut adjustment against locking force for the actual machine being used.

However, suppose graphs are available of the relationships involved. Assume that the machine is of 200 tons and the tie bar nuts can be adjusted to ~½ turn from maximum lock, and that a further division is backed off for a thermal expansion allowance. It can be seen from the graphs that 1½ turns would give a locking force of ~175 tons. Similarly, if instead of working at 2000 lb/in^2 it is assumed the locking pressure is ~1800 lb/in^2, it is found that this will lower the lock by ~20 tons. Thus the above allowance will produce a working locking force of ~155 tons.

Allowance for Pressure Peaks and Intensification

This is the most difficult factor to quantify, because machines can vary in their design and hence in the inertia effects they impart to the metal in the die. Similarly, intensifiers can be set to vary as to when they have an effect.

However, multiplication factors can be used to give working approximations. The following are suggested:

	× Factor
Hot chamber zinc machines with heavy pneumatic pistons	2.5 P
Hot chamber zinc with hydraulic injection	1.5 P
Older cold chamber aluminium machines with accumulators at the locking end	2.0 P
Modern cold chamber aluminium machines	1.25P

The above factors make no allowance for the increased penetration at the parting line or at sliding cores when high injection speeds are used, nor does it take into account whether the intensification pressure is applied early or late in the casting cycle. Metal fluidity is also a difficult factor to qualify, but a fluidity factor of 75% is suggested, i.e. it is assumed that only 75% of the intensification pressure is applied to the die cavity.

REFERENCES

1. A. J. Davis, The injection process in Diecasting. 9th International Pressure Diecasting Conference, London, 1978.
2. S. E. Booth and D. F. Allsop, BNF Manual of Pressure *Diecasting Gating Design.* BNF Metals Technology Centre, Grove Laboratories, Denchworth Road, Wantage, Oxon.
3. "Gateway" locking, gating and thermal die design system for diecasting. BNF Metals Technology Centre, Grove Laboratories, Denchworth Road, Wantage, Oxon.
4. A. J. Wall and D. L. Cocks, Diecasting technology. Developments in zinc. 77th IBF Conference, *British Foundryman* Oct. 1980, vol. 73.
5. S. E. Booth, *BNF Guide to Better Aluminium Diecasting.* BNF Metals Technology Centre, Grove Laboratories, Denchworth Road, Wantage, Oxon.
6. J. W. O'Brien, Method of measuring tie bar stress in diecasting machines, *Trans. SDCE,* 1977, paper GT 77-045.
7. T. B. Hill, Private communication.

Chapter 4

Furnaces Used in Pressure Diecasting and the Assessment of their Performance

Generally speaking, furnaces used in diecasting do not pose great technological problems. It is taken for granted that they will produce metal of the required quantity and quality and until quite recently they were not often critically examined. This attitude is now changing, particularly with regard to the types of heating being used, its efficient use, the types of elements, burners and refractories available, the use of recuperators, etc. There is likely to be an even greater economic advantage as fuel prices rise, and to obtain the facts on furnace performance, etc., at shop floor level will be the responsibility of a technician.

This chapter is intended to give a background to furnaces used in diecasting and to demonstrate some basic calculations from which a useful indication of furnace performance can be assessed.

1. TYPES OF FURNACES

There are numerous types and sizes of furnaces available to diecasters. Details can readily be obtained from furnace maker's literature, and there is no point in repeating what is readily available from these sources. However, a brief summary of the furnace position is worth stating if only to simplify what may at first sight appear a complicated situation.

If we first consider fossil fuel furnaces, i.e. those using oil and gas. These can be divided into two broad categories, namely reverberatory furnaces, where the flame and combustion products are in contact with the metal being melted, and crucible furnaces,

84

where there is no such contact. In diecasting, reverberatory furnaces are most frequently used for the bulk melting of ingots and scrap or, in a modified form, for the reception and holding of bulk liquid metal received by road or rail transport. After melting or holding, molten metal is then transferred to smaller furnaces at the diecasting machines. The furnaces themselves are simply steel shells lined with refractory brick or having a monolithic lining. A burner is fixed to one end wall and the exhaust stack at the other. Charging is carried out either by a separate door or through the exhaust stack. Metal is usually removed by either tilting the furnace or through a tap hole.

A variant of the reverberatory furnace is the radiant furnace. Here the burners are mounted in the roof rather than on the end wall, and hence the chances of direct flame impingement on the metal is greatly reduced, but the advantages of radiant heat from the flame and refractories is maintained.

Crucible furnaces are generally much smaller than reverberatory units, and are commonly used as holding units at each diecasting machine. In their simplest form they consist of a steel case lined with refractory brick into which a crucible is placed. The gap between the crucible and refractory is the combustion space. The furnace is completed by an exhaust flue at the top side and a burner, usually firing tangentially into the bottom. Metal is transferred from the holding units to the diecasting machine either by a hand ladle or automatic transfer system.

Despite their general use as holding units, many furnaces of the above type are used as melting units, and thus serve a dual role. Ingot and scrap is charged at the discretion of the diecaster to balance the metal cast. A variation on the crucible furnace is the melting/holding unit of a hot chamber diecasting machine. Here the crucible is of cast iron heated by burners from below, which are in turn enclosed in a steel box acting as a combustion space.

Larger crucible furnaces also find use as bulk melting units, and here they are invariably made to tilt in line with the pouring spout to facilitate metal transfer.

There are two other types of furnace which do not fall into either of the above categories, but which are worth mentioning. These are the immersed crucible (IC) furnace and the immersion tube furnace.

The principle of the IC furnace is illustrated in Fig. 4.1[1]* where it will be seen that the burner is fired into a crucible rather than outside it; heat is conducted through the crucible and also radiated from the

*Superscript numbers refer to References at end of chapter

flame itself, and the roof refractories heated by the flame as it passes to the exhaust stack. This type of furnace is used both as a bulk melting unit to serve smaller bale-out furnaces or as a furnace to serve one diecasting machine. As a bulk melter, metal is removed by a tap-out valve below metal level, but if attached to one machine, metal is transferred as if it were a crucible furnace by removing part of the refractory lid. The immersion tube furnace is shown in Fig. 4.2.[2] Here the metal is heated in a refractory lined steel bath by several stainless steel tubes through which are passed the products of combustion from burners mounted at one end. Metal removal is achieved by a tap hole. The furnace is used as a melting or holding unit, but is obviously limited to metals which are non-aggressive to the steel from which the tubes are made.

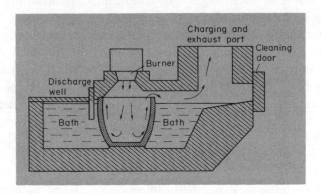

FIG. 4.1 Diagram of Immersed Crucible Furnace

This is a short summary of the fossil-fired furnace position, but before leaving the subject of furnace types it is as well if we consider electric heating.

Electric heating has in the past been less common in the United Kingdom due to capital and running costs being higher than the alternatives. However, the increasing cost of fossil fuel and, in some instances, the inability of companies to purchase much more than their present demand, has caused a renewed interest in electric heating. Electric furnaces fall into two categories, namely induction furnaces and resistance furnaces.

Induction furnaces themselves fall into two categories, the channel induction furnace and the coreless induction furnace.

The channel induction furnace consists of two main parts, the holding bath and the inductor box (Fig. 4.3). Electrically the furnace can be regarded as a transformer in which the alternating power is applied to a primary coil wound on a laminated iron core, whilst the secondary is the channel which contains a portion of the furnace charge and is connected to the main charge at both ends. A high current is induced in the secondary, i.e. the metal in the channel and the hot metal is pushed by convection currents into the main charge in the body of the furnace. This is replaced by colder metal and the cycle is repeated.

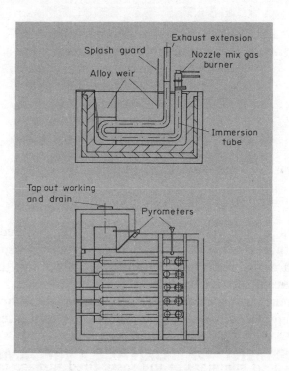

FIG. 4.2 DIAGRAM OF IMMERSION TUBE FURNACE

Channel furnaces can be used for melting and holding purposes.

The other type of induction furnace is the coreless induction furnace. This usually consists of a multiple-turn water-cooled coil surrounding a crucible or rammed lining. The frequency used can vary, but irrespective of frequency, a powerful stirring action is imparted to the melt. Variations of the induction furnace are the crucible and coil being quickly separated from each other, as in the lift-off coil arrangement or where the crucible itself is pushed up through the coil by a hydraulic cylinder. Both these arrangements eliminate the need for a tilting mechanism on the furnace, since, if small enough, the crucibles can easily be transported.

Resistance furnaces also fall into two types, those which radiate heat directly onto the furnace charge and those which radiate heat onto a crucible. The first category are in effect reverberatory furnaces similar to the radiant type, whilst the second category are similar to the crucible furnaces. Both types have been described previously.

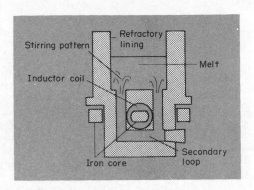

Fig. 4.3 Diagram of Channel Induction Furnace

A variant of the resistance furnace is the so-called low-energy holding furnace. These can vary in their detailed design, but essentially they are steel shells lined with refractories of low thermal mass; the heating elements are in the roof.

These units are uncompromisingly designed as holding units, melting being virtually impossible except over very extended times.

2. ASSESSMENT OF FURNACE PERFORMANCE

Equipment Needed

Since the purpose of any furnace is to melt or hold metal using a minimum of energy, the performance of a furnace is a measure of the amount of energy used to hold or melt a given amount of metal. To assess this we have to take certain measurements which are as follows:

1. Metal temperature.
2. Weight of metal charge.
3. Energy usage.

In addition it is necessary to calculate the heat losses from the furnace, i.e. the heat not going into the metal. For this calculation the following data is required:

4. Furnace case temperature.

5. Stack gas temperature.
6. Stack gas analysis.

To measure these the following equipment is necessary.

Meters

For oil, these can be simple flow meters fitted into the oil feed line, giving a direct reading in gallons or litres. For gas, a typical displacement gas meter can be used giving a direct reading in cubic feet (kWh meter for electricity).

An alternative to the above for oil or gas is the turbine meter (Fig. 4.4). A tee-piece is connected into the oil or gas line and the turbine head inserted into the centre of the pipe. Movement of oil or gas turns the turbine and the number of turns is counted on an electronic meter. The number of counts is proportional to fuel flow.

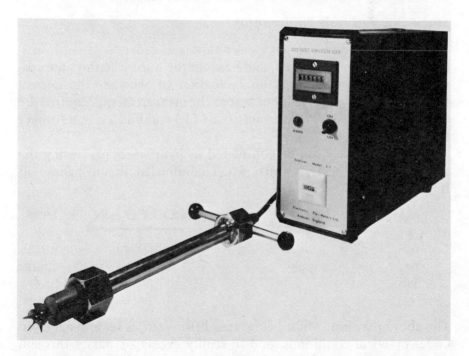

FIG. 4.4 TURBINE MEASURED HEAD AND ELECTRONIC COUNTING UNIT

This method is useful when the fuel flow figures are processed further, as in a microprocessor. This is not possible using the simple flow meters, which have to be read manually. These are nevertheless relatively inexpensive and adequate for many purposes.

Pyrometers

For measurement of stack gas temperature a pyrometer capable of reading 200 - 300°C above the casting temperature of the alloy in the furnace is needed. Thus to cover copper base alloys ~1500°C maximum is required, whilst for zinc, ~650°C is probably satisfactory. Care must be taken when measurements are taken to avoid flame impingement, which can give a falsely high reading. The preferred method which can overcome this problem is to suck a portion of the stack gases out through a small tube inserted into the stack and pass them over a thermocouple; flame impingement is then avoided. This method can also provide the stack gas sample for analysis. To measure furnace case temperature, either the press contact type or limpet instruments are used, whilst for metal temperature the standard type of thermocouple used in the foundry is adequate.

Stack Gas Analysis Equipment

We need to know the stack gas analysis in order to see how efficiently the fuel is being used, and hence if the burner settings are correct. Stack gas analysis is an indication of whether the correct fuel/air ratio is being used, but before the measurements required are discussed it is as well if the chemistry of fuel combustion in a furnace is considered.

Oil and gas fuels can be considered as hydrocarbons, having the general formula C_xH_x, which when combusted in air behave as follows:

$$C_xH_x \quad + \quad \underbrace{O_2 + N_2} \quad \longrightarrow \quad \underbrace{H_2O + CO_2 + N_2} \quad + \quad \text{Heat}$$

oil	air	stack gas	metal
or	supply		charge
gas			+
			losses

The above reaction, whilst being straightforward, is seldom achieved under practical conditions, and in reality excess air passes through the burner which will give excess O_2 in the stack gases, or alternatively excess fuel which can give rise to CO and H_2 in the stack is present, since there will be insufficient O_2 to give CO_2 or H_2O. In extreme cases of excess oil fuel, soot which is unburned carbon, is present. The above chemical reaction, therefore, can only take place at the exact ratio of fuel to air, and this is called the stoichiometric composition. This is the ideal condition to operate a furnace from the fuel efficiency point of view. However, for metallurgical reasons, particularly in reverberatory type furnaces, an oxidising atmosphere must sometimes be avoided and in practice this will mean running the furnace with marginally excess fuel. A saving in fuel is not economic if the metal quality suffers as a result. It should be noted, that N_2 in the incoming air supply is not involved in the combustion of fuel. It is nevertheless heated by the fuel, and this represents a serious loss of heat.

From the above, it follows that the combustion of fuel can be followed by an analysis of the stack gases and Fig. 4.5 shows the relationships involved. If stoichiometric combustion is considered, there is a maximum of ~12% CO_2 and zero O_2, H_2 and CO. If excess air is present, the CO_2 content will be lower and O_2 will be present in amounts depending upon the excess air level. If excess fuel is used, CO_2 will again be lower and O_2 will be zero, but in this case CO or H_2 will be present in the stack gases. In practice, the measurement of O_2 and CO_2 is usually sufficient to assess efficiency with sometimes an occasional check for CO.

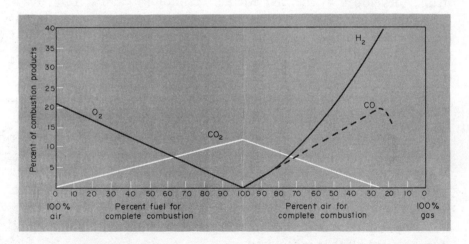

FIG. 4.5 PERCENTAGE OF COMBUSTION PRODUCTS AGAINST FUEL-AIR RATIO

The equipment used for analysis falls into two basic types, chemical analysers and electronic methods. The two common chemical methods are the "Orsat" and the "Fyrite". With such analysers, the flue gas is sampled via a steel tube positioned in the stack, and piped by a rubber hand pump into the analyser vessel. The method is to absorb selectively the gas being analysed, and its composition is registered by a change in volume of the liquid and read out directly from a calibrated scale. The method is cheap, rapid and sufficiently accurate for most purposes. It is, however, a manual method and the results cannot be fed automatically into a microprocessor such as that used as part of a feedback circuit for burner control.

The electronic methods have the advantage of being integrated into electronic control systems and they can be used for isolated or continuous sampling. One analyser can effectively monitor and provide the necessary control data for many furnaces without the necessity of an operator. The systems themselves can be based on different principles.

One system passes the stack gases into a small furnace where it comes into contact with a ceramic oxygen sensor which produces a voltage proportional to the oxygen level. Another system measures CO_2 indirectly by thermal conductivity measurements of the stack gas, whilst another method measures the voltage change in an electrolyte when it is combined with O_2.

An example of an electronic system is the furnace diagnostic trolley developed by the BNF Metals Technology Centre[3] (Fig. 4.6). This system uses a microprocessor to monitor the instruments fitted to furnaces and then uses this information for burner adjustment. The instrument monitors signals from gas analysers, flow meters, and thermocouples. It then calculates the stack heat losses, furnace efficiency and the optimim fuel/air ratio. The calculations are displayed, with indications of how the fuel/air ratio should be changed to optimise combustion conditions. The system can be used at a central point to monitor and control many furnaces simultaneously.

Even without the use of sophisticated measuring equipment the same results can be achieved by arithmetic and the use of tables, graphs and nomograms. The calculations take longer and the effect of varying experimental conditions will take longer to check, but the results obtained are as accurate and for many applications involving furnaces are all that is required.

Furnace Calculations

From the list given previously, there are two measurements more important than others, namely weight of metal in the charge, and fuel used to melt the charge or to hold it at a given temperature. The ratio of fuel used to the amount of metal melted or held is a measure of the efficiency of the furnace and can be used as a basis for comparison between different types of furnace and fuels. The other

Fig. 4.6 BNF Metals Technology Centre Furnace Diagnostic Trolley

measurements are not fundamental to the assessment of furnace efficiency but they are required to assess where losses are occurring in the operation of a furnace and to reduce them.

FURNACE EFFICIENCY

It is necessary to define what is meant by furnace efficiency and to discuss and describe other terminology in greater detail. If fuel is considered, it is normal in the United Kingdom to find that oil is quoted in gallons or litres, gas in cubic feet or British thermal units, whilst electricity is quoted in kilowatt hours. Similarly, metal weight is quoted in kilogrammes or tonnes. What is needed, therefore, is to use one unit of heat measurement to cover all fuels and this can be either the joule, calorie, therm or kilowatt. In the United Kingdom the most common standard unit of heat in practical furnace calculations is the therm, whilst the standard unit of metal weight is the ton. Thus when the amount of heat required to melt a particular alloy is discussed, therms/ton are used. Figure 4.7 shows the heat content of lead, zinc, aluminium, magnesium and copper at various temperatures in terms of therms/ton as well as in other terminology, where it will be seen that at typical diecasting temperatures the heat content of the various metals are as follows:

Lead	~ 0.75 therms/ton
Zinc	~ 2.75 therms/ton
Aluminium	~ 11.00 therms/ton
Magnesium	~ 11.00 therms/ton
Copper	~ 7.00 therms/ton

The above are for the metal elements, but in practice we make little error if we use the same values for alloys based on these elements. The above values, therefore, represent the 100% efficiency condition, i.e. if we have 1 ton of aluminium and only require 11 therms to heat it to casting temperature, this represents 100% efficiency. In real life, however, this is never attained. For aluminium, values in the range of 50 - 200 therms/ton are more common, and these represent furnace efficiences of:

$$\frac{11}{50} \times 100 = 22\% \text{ efficiency}$$

and

$$\frac{11}{200} \times 100 = 5.5\% \text{ efficiency}$$

Thus furnace efficiency expressed as a percentage is seen to be the ratio of the theoretical amount of heat in therms/ton to the actual amount of heat used in therms/ton × 100.

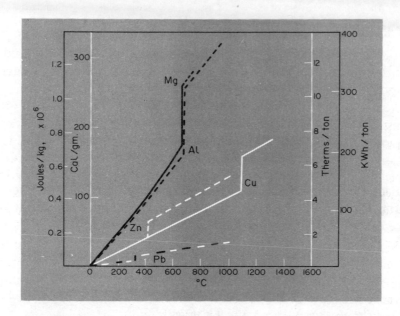

FIG. 4.7 HEAT CONTENT OF METALS BY WEIGHT (SPIEN)

For zinc a real figure of 15 therms/ton would be:

$$\frac{2.75}{15} \times 100 = 18.3\% \text{ efficiency}$$

whilst for copper base alloys a figure of 75 therms/ton would be:

$$\frac{7}{75} \times 100 = 9.3\% \text{ efficiency}$$

There are, however, pitfalls in using a standard value of therms/ton irrespective of the alloy casting temperature and, strictly speaking, the therms/ton value for the actual casting temperature of the alloy should be used. This is well illustrated by a case history. The records of a 6-ton electric furnace were being analysed and the results obtained are shown in Table 4.1.

On the basis of 7 therms/ton for copper base alloys, the furnace efficiencies would be 25.3%, 24% and 21.3% respectively. However, taking into account the theoretical therms/ton at the casting temperatures, being ~6.5, 7 and 7.5, the corrected furnace efficiency values were 23.5%, 24% and 22.9%. Having seen how to arrive at a

TABLE 4.1 MEASUREMENTS ON A 6-TON ELECTRIC FURNACE

Alloy	No of melts	Casting Temp. °C	Mean value, Therms/ton	Standard deviation (σ)
Manganese bronze	19	1070	27.7	3.4
Copper-manganese - Aluminium	30	1190	29.2	3.4
Aluminium bronze	18	1270	32.8	5.6

furnace efficiency figure it is necessary to see how the number of therms is determined to melt the metal in question. This can be obtained from the conversion factors set out in Table. 4.2.

TABLE 4.2 CONVERSION FACTORS FOR VARIOUS FUELS AND WEIGHTS

1 gallon of gas oil	= 1.64 therms
1 litre of gas oil	= 0.36 therms
29 kWh	= 1.00 therms
97 ft^3 natural gas	= 1.00 therms
1 gallon	= 4.55 litres
1 ton	= 1016 kg

Using Table 4.2, for a furnace which requires 379 litres of gas oil to melt 1.75 tons of aluminium alloy the efficiency is:

379 litres $\times$ 0.36 = 136.4 therms to melt

1.75 tons $= \dfrac{136.4}{1.75} = 77.9$ therms/tons.

% efficiency $= \dfrac{11}{77.9} \times 100 = 14.1\%$

Similarly, for a furnace which requires 4842 ft^3 of natural gas to melt 1.25 tons of copper alloy the efficiency is:

4842 ft^3 $\times \dfrac{1}{97} = 49.9$ therms to melt

1.25 tons $= \dfrac{49.9}{1.25} = 39.9$ therms/ton.

% efficiency $= \dfrac{7}{39.9} \times 100 = 17.5\%$

In practice, furnace efficiency is not as clearcut a yardstick as the above calculations would suggest. In fact, there is no single value of effciency which can be given to a furnace even when melting the same alloy. Furnace efficiency for any one furnace and burner may depend upon several factors, some of which are listed below.

1. Actual charge weight and its physical condition.
2. Accuracy of burner settings.
3. Melting time.
4. Condition of furnace lining.
5. Whether first heat or subsequent heats.

The sort of results obtained in real life are detailed in Table 4.3, which sets out an actual furnace log for 1 day's operation. The furnace was a reverberatory oil-fired unit melting copper base alloys for sand castings, but similar variations would be obtained on other furnaces and alloys. The lowest efficiencies are associated with small charge weights (heat 3), i.e. a partly filled furnace. It is well known that a furnace will melt a full load for very little more fuel than a partial load. This fact emphsises the importance of the correct furnace size relative to metal requirements. Another more obvious case of lower efficiency is the heat absorbed by the cold refractories at start up (heat 1).

The furnace log indicates the amount of heat used holding the furnace at temperature during the various other foundry operations such as shanking the metal around the foundry and illustrates that furnace efficiency is very much dependent upon foundry practice.

On the day in question, metal was being melted in the furnace with an average efficiency of 17.23%, but the fuel used during the other operations caused this to fall to an average of 13.55%.

So far the furnace has been considered as a melting unit. There is, therefore, the holding situation to consider. If a holding furnace is constantly being refilled with molten alloy to replace the metal cast, the heat supplied by the furnace need only be sufficient to balance out the radiation and convection losses from the furnace walls and metal surface, and this is only a proportion of the heat needed to melt the alloy in the first instance. Under these conditions efficiencies cannot be calculated, since there is no theoretical value of 100% as there is in melting and hence a holding furnace performance is normally quoted in therms/hr. This is not, however, an accurate basis for comparison unless furnaces are of the same physical size and holding capacity. It is obvious that a larger furnace will hold more metal than a smaller one and because of its greater surface area

TABLE 4.3 FURNACE LOG FOR REVERBERATORY FURNACE MELTING
COPPER BASE ALLOYS

Charge no.	Time	Oil meter reading, litres	Remarks
1	8.43	25,180.68	681 kg phosphor bronze charged. Stack gas analysis 13% CO_2
	10.50	25,282.60	Casting started. Furnace case temperture 90°C.
	11.19	25,303.92	Casting completed.
2	11.21	25,303.92	681 kg BS 1400 LB3 charged.
	12.45	25,363.20	Casting started. Furnace case temperature 240°C.
	1305	25,378.7	Casting completed.
3	13.28	25,378.7	100 kg BS 1400 PB2 charged.
	13.50	25,397.4	Casting started. Furnace case temperature 240°C
	13.53	25,397.4	Casting completed.
4	13.56	25,397.4	545 kg BS 1400 LG4 charged.
	14.56	25,439.8	Metal ready to be cast.
	15.13	25,453.6	Casting started. Furnace case temperature 240°C.
	15.25	25,463.3	Casting completed.

Calculation of furnace efficiency based on Cu base alloys at 7 therms/ton

Charge 1

Fuel used in melting = 25,282.60 - 25,180.68 = 101.92 litres
= 36.69 therms

Metal melted = 681 kg = 0.669 tons

Therms/ton = $\frac{36.69}{0.669}$ = 54.84

% efficiency = $\frac{7}{54.84}$ × 100 = 12.76%

Overall fuel used in melting and casting
= 25,303.92 - 25,180.68 = 123.24 litres = 44.36 therms

Therms/ton = $\frac{44.36}{0.669}$ = 66.3

% efficiency = $\frac{7}{66.3}$ × 100 = 10.56%

Charge 2

Fuel used in melting = 25,363.20 - 25,303.92 = 59.28 litres
= 21.34 therms

TABLE 4.3 *(cont.)*

Metal melted = 681 kg = 0.669 tons

Therms/ton = $\dfrac{21.34}{0.669}$ = 31.89

% efficiency = $\dfrac{7}{31.89}$ × 100 = 21.95%

Overall fuel used in melting and casting
= 25,378.7 - 25,303.92 = 74.78 litres = 26.92 therms

Therms/ton = $\dfrac{26.92}{0.669}$ = 40.24

% efficiency overall = $\dfrac{7}{40.24}$ × 100 = 17.39%

Charge 3

Fuel used in melting = 25,397.4 - 25,378.7 = 18.7 litres
$\qquad\qquad\qquad\qquad\qquad\qquad\qquad\qquad$ = 6.73 therms

Metal melted = 100 kg = 0.0982 tons

Therms/ton = $\dfrac{6.73}{0.0982}$ = 68.53

% efficiency = $\dfrac{7}{68.53}$ × 100 = 10.2%

No fuel was used to heat the metal during casting, so the overall efficiency is 10.2%.

Charge 4

Fuel used in melting = 25,439.8 - 25,397.4 = 42.4 litres
$\qquad\qquad\qquad\qquad\qquad\qquad\qquad\qquad$ = 15.26 therms

Metal melted = 545 kg = 0.535 tons

Therms/ton = $\dfrac{15.26}{0.535}$ = 28.54

% efficiency = $\dfrac{7}{28.54}$ × 100 = 24.52%

Fuel used "holding" furnace due to foundry, problems
= 25,453.6 - 25,439.8 = 13.8 litres = 4.97 therms

Therms/ton = $\dfrac{20.23}{0.535}$ = 37.81

% efficiency = $\dfrac{7}{37.81}$ × 100 = 18.51%

Overall fuel used in melting, holding and casting
= 25,463.3 - 25,397.4 = 65.9 litres = 23.72 therms

Therms/ton = $\dfrac{23.72}{0.535}$ = 44.34

% efficiency overall = $\dfrac{7}{44.34}$ × 100 = 15.79%

TABLE 4.3 *(cont.)*

Summary of performance

Overall fuel used for four heats, melting only = 222.3 litres
= 80.03 therms

Overall metal melted = 2007 kg = 1.97 tons

Therms/ton = $\dfrac{80.03}{1.97}$ = 40.62

% efficiency = $\dfrac{7}{40.62}$ × 100 = 17.23%

Overall fuel used in melting, holding and casting = 282.6 litres = 101.74 therms

Therms/ton = $\dfrac{101.74}{1.97}$ = 51.64

% efficiency overall = $\dfrac{7}{51.64}$ × 100 = 13.55%

will have more losses and hence need a larger burner and consume more fuel. If a simple therms/hr is taken as the basis of comparison, the large furnace will perform unfavourably relative to the small one, and under these circumstances a better comparison is obtained by dividing the therms/hr by the furnace tonnage to give therms/ton/hr.

In addition to a furnace being used as a melting unit or a holding unit, there is the possibility that it is used as a combination of both. This situation is common for the immersed crucible and the melter bale-out furnaces. To check such situations it is necessary to start and finish the checking exercise with the same conditions in the furnace.

If an immersed crucible furnace initially holding 1 ton of molten metal is considered, the amount of fuel required to melt the particular throughput of ingot is recorded. It is still necessary to finish up with 1 ton of molten metal in the furnace at the end of the exercise, i.e. if the throughput in a particular time is 2 tons, the efficiency of the furnace when melting 2 tons of metal is being measured.

Any other method of checking a furnace working than that described above can lead to the wrong conclusions. Generally speaking it is sufficiently accurate to ascribe the amount of fuel used to the melting of the alloy, since this will be much more than the fuel used to maintain the molten alloy at temperature, but if a more accurate assessment is needed, the fuel used in holding must be taken into account. If we only take account of the metal melted and the fuel used, furnace efficiency is calculated as previously described.

Given the various factors which can influence furnace efficiency that are outside the control of the furnace builder, it is not surprising that many furnaces are sold without even a mention of efficiency, and this is not surprising, since to quote a figure may lead foundries to expect it to be met irrespective of how the furnace is operated. However, some furnace catalogues do quote figures of fuel consumption, melting times and furnace capacities from which efficiencies can be calculated using the values given in Table 4.2.

The type of information typically quoted by furnace builders is shown in Table 4.4.[1] Table 4.5 is the same data as Table 4.4 but instead of quoting melting times, weights and therms used, the values quoted in Table 4.4 have been used to calculate the data given in Table 4.5 from which a better comparison of furnace performance can be obtained.

TABLE 4.4 TYPICAL PERFORMANCE DATA FOR BALE-O-MATIC III
FURNACE WHEN MELTING OR HOLDING ALUMINIUM

Capacity	kg	50	65	100	135	175**	250	300	350	500
	lb	100	140	200	300	400	550	660	800	1200
Melting time	First heat min	90	100	130	130	140	170	195	225	280
Aluminium at	Subsequent									
720°C	heat min	65	70	90	90	100	120	140	160	210
Fuel consumption	kcal × 10³	145	145	155	155	205	205	225	265	310
per hour	therms	5.74	5.74	6.23	6.23	8.2	8.2	9.00	10.65	12.30
Melting at maximum										
throughput	litres	16	16	17	17	23	23	25	29	34
Aluminium at 720°C	gallons	3.5	3.5	3.8	3.8	5.0	5.0	5.5	6.5	7.5
Fuel consumption	kcal × 10²	16	16	28	28	47	47	50	50	64
per hour	therms	0.66	0.66	1.15	1.15	1.90	1.90	2.00	2.00	2.50
maintaining only										
Aluminium at 720°C	litres	1.80	1.80	3.20	3.20	5.20	5.20	5.40	5.40	7.00
	gallons	0.40	0.40	0.70	0.70	1.15	1.15	1.20	1.20	1.55
Melting maximum										
throughput	kg/hr	55	65	70	90	130	145	180	210	240
Aluminium at 720°C	lb/hr	125	140	150	200	280	320	380	450	510

**With BC202 crucible

Since many diecasters operate with a number of alloys, it is a useful exercise to compare their respective melting costs.

Figure 4.8 is a graph of melting costs/ton for zinc and aluminium alloys for various furnace efficiencies quoted both as percentage efficiency and in therms/ton. It is obvious from the graph that for the same efficiency, zinc is much cheaper to melt than aluminium. For example, at 10% efficiency, zinc, melting cost is ~ £9/ton and aluminium $\sim$£33/ton with fuel at 30p/therm (1980). However it must be remembered that 1 ton of aluminium will produce a larger number of castings than 1 ton of zinc. The density of aluminium is $\sim$2.8 g/cm^3 and zinc is $\sim$6.7 g/cm^3, therefore 1 ton of aluminium will produce $\sim$2.4 times as many identical castings at 1 ton of zinc.

If we compare the melting cost/casting using zinc, aluminium, magnesium and copper, we have the situation shown in Table 4.6 and in Fig. 4.9 If a casting made in zinc is assumed to weigh 0.5 lb (163 cm^3 volume), and hence 4480 castings/ton, the number of castings/ton in the other alloys is obtained from the ratio of the alloy density to that of zinc. Since the value in therms/ton to melt the alloys and the cost of fuel/therm is known, a group can be constructed of melting efficiency against melting cost/casting as shown in Fig. 4.9. If the melting cost/casting is taken in zinc as equal to 1.0, the melting cost/casting for the other alloys can be calculated, and these are shown in the end column of Table 4.6.

Zinc is still the cheapest alloy, but only marginally better than magnesium and, in practice, there would probably be no difference. Copper base alloys are three times as expensive as zinc and aluminium $\sim$1.7 times as expensive.

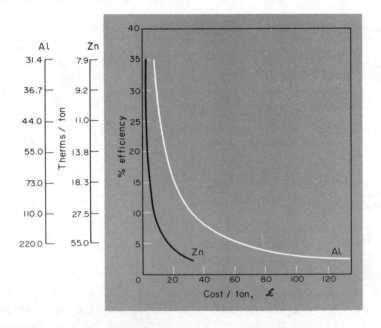

Fig. 4.8 Melting Cost/ton in £ of Zn and Al at 30p/therm

TABLE 4.5 TYPICAL PERFORMANCE DATA IN THERMS/TON AND % EFFICIENCY FOR BALE-O-MATIC III FURNACE WHEN MELTING OR HOLDING ALUMINIUM

Furnace Capacity, lb	Specification Capacity, tons	Melting efficiency 1st heat		Melting efficiency Subsequent heats		Efficiency at max. melting rate (i.e. melting & casting)		Holding	
		therms/ton	%	therms/ton	%	therms/ton	%	therms/hr	therms/ton/hr
	Aluminium to 720°C								
100	0.045	191	5.7	138	8.0	104	10.6	.66	14.7
140	0.063	152	7.2	106	10.4	92	12.0	.66	10.5
200	0.089	152	7.2	105	10.5	93	11.8	1.15	12.9
300	0.134	101	10.9	70	15.7	70	15.7	1.15	8.6
400	0.179	107	10.3	76	14.5	66	16.7	1.90	10.6
550	0.246	94	11.7	67	16.4	57	19.3	1.90	7.7
660	0.295	99	11.1	71	15.5	53	20.8	2.00	6.8
800	0.357	112	9.8	80	13.8	53	20.8	2.00	5.6
1200	0.536	107	10.3	80	13.8	54	20.4	2.50	4.7

TABLE 4.6 RELATIVE MELTING COST FOR A CASTING OF 164 CM^3 VOLUME

Alloy	Density gas/cm^3	Casting wt. lb	Castings /ton	Relative melting cost /casting
Zn	6.7	0.5	4480	1.0
Mg	1.8	0.13	17,231	1.04
Cu	8.3	0.62	3613	3.16
Al	2.8	0.21	10,667	1.68

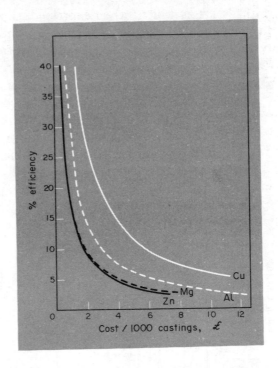

FIG. 4.9 MELTING COST/1000 CASTINGS OF VOLUME 164 CM^3 AT 30P/THERM

Figure 4.9 also shows the efficiencies needed in the higher cost aluminium and copper alloys to be able to achieve the same melting costs/casting as zinc and magnesium, e.g. if zinc is melted at 5% efficiency (which is not an untypical figure) it is necessary to melt aluminium at ~ 9% efficiency to achieve the same melting cost/casting.

The above figures do not take into account casting yield, i.e. the ratio of casting weight to shot weight, and this will vary depending on the casting. Similarly, an allowance for metal losses is not included.

Irrespective of whether the fuel used/ton or fuel used/casting is considered, the fact that the majority of diecasting foundaries operate at between ~5 and 15% efficiency means that there is great scope for improvement.

FURNACE LOSSES

From practical experience it is obvious that heat is emitted from the furnace walls, the metal surface and the exhaust stack gases. In addition, heat is also lost to the furnace refractories bringing the furnace to its equilibrium working temperature.

These constitute losses which, when expressed as percentages and added to the percent efficiency of the furnace, should add up to 100%, equivalent to the numbers of therms in the fuel used for the particular furnace charge. In practice is is unusual to be able to account for 100% of the energy used. This is due to such factors as inaccuracy in measuring stack gas analysis and temperature (furnace case temperatures are not the same at all points on the case), air leaks into the furnace, additional losses from open charging doors, etc. Generally it is sufficiently accurate if furnace efficiency, metal surfaces losses, and furnace case losses are calculated. If these added together and subtracted from 100% give a figure approaching the value of stack gas losses obtained from the nomogram (Fig. 4.10), there is little reason to be concerned.

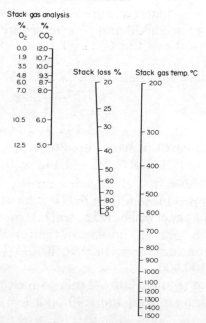

Fig. 4.10 Nomogram for Determining Stack Losses from Stack Gas Temperature and Analysis

Furnace Case Losses

Convection and radiation losses are estimated from the furnace case temperature, the area of the furnace walls (not forgetting the furnace base and furnace top) and the values given in Fig. 4.11.

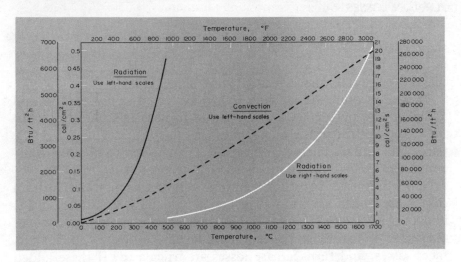

FIG. 4.11 HEAT LOSS FROM SURFACES BY RADIATION AND CONVECTION

For example, a furnace 5 ft in diameter and 4 ft in height containing a crucible 1.5 ft in diameter with a furnace wall temperature of 200°C melts a charge in 2 hr 20 min. The losses are calculated by the following steps.

1. Area of furnace side wall $= \pi D \times H$
   $$= 3.142 \times 5 \times 4 = 62.84 \text{ ft}^2$$
2. Area of base $= \pi r^2$
   $$= 3.142 \times 2.5^2 = 19.64 \text{ ft}^2$$
3. Area of top = area of base - area of crucible
   $$= (3.142 \times 2.5^2) - (3.142 \times 0.75^2) = 17.87 \text{ ft}^2$$
4. Total surface area $= 100.35 \text{ ft}^2$
5. Radiation losses at 200°C (Fig. 4.11) $= \sim 900 \text{ BTU/ft}^2\text{/hr}$
6. Convection losses at 200°C (Fig. 4.11) $= \sim 500 \text{ BTU/ft}^2\text{/hr}$
7. Total losses $= \sim 1400 \text{ BTU/ft}^2\text{/hr}$
8. Total losses on furnace $= 1400 \times 100.35 = 140490 \text{ BTU/hr}$
9. Dividing by 100,000 $= 1.40 \text{ therms/hr}$
10. If the furnace requires 2 hr 20 min to melt the charge, the losses during melting are $1.40 \times 2.3 = 3.22 \text{ therms}$

Losses from Metal Surface

Convection and radiation losses are estimated from the surface area of the metal and its temperature.

For example, if the same furnace as before, is considered with a crucible of 1.5 ft diameter containing aluminium at 750°C, the losses are obtained by the following steps.

1. Area of metal surface = πr^2
    $$= 3.142 \times 0.75^2 = 1.77 \text{ ft}^2$$
2. Radiation losses at 750°C (Fig. 4.11)   = ~ 20,000 BTU/ft²/hr
3. Convection losses at 750°C (Fig. 4.11) =   ~ 2400 BTU/ft²/hr
4. Total losses   = ~ 22,400 BTU/ft²/hr
5. Total losses from metal surface = 22,400 × 1.77
    $$= 39,648 \text{ BTU/hr}$$
6. Dividing by 100,000   = 0.3965 therms/hr
7. If the furnace requires 2 hr 20 min to melt the charge, the losses during melting are 0.3965 × 2.3 = 0.91 therms.

Pitfalls in Calculations

The above calculations are given as examples, and are not strictly valid in some instances for the following reasons.

1. The furnace case temperature will not be constant at 200°C for the whole of the charge time. If the furnace was initially at a lower temperature, it may be more accurate to estimate losses based on the average wall temperature during the charge.
2. The losses from the metal surface assume the metal to be at 750°C, which is only the case when the metal is molten and not during the heating-up period. However, when melting, a heel of liquid metal is quickly formed, so this assumption may not be too inaccurate. The calculation is, however, completely valid for a holding furnace.

Stack Gas Losses

These are most easily assessed using a nomogram (Fig. 4.10) which requires a knowledge of the CO_2 in the stack gas and the stack gas temperature. A line is drawn joining the percentage of CO_2 to the temperature; where it crosses the central scale the stack loss can be read off directly.

3. SOME ASPECTS OF IMPROVING FURNACE PERFORMANCE

Since data has been quoted in terms of percentage, it may be thought that the ultimate in furnace efficiency should approach 100%. In practice this is not the case, since there are some losses which cannot be eliminated,[4, 5] mainly stack gas losses. In any fossil fuel furnace the metal is heated by the products of combustion, which impart heat to the charge and escape through the stack. Heat can only be transfered if the gases are at a higher temperature than the metal, and in practice the gases may be at 100-200°C higher temperature than the metal, thus the heat retained by these gases is the major source of thermal loss, which is why losses are higher with the high melting-point metals. The relationship between flue gas temperature and percentage of gross fuel input which is available as possible useful heat is shown on Table 4.7. It will be seen that even with perfect combustion conditions the ultimate for

Zinc is   65 - 70%
Aluminium and magnesium is   50 - 55%
Copper base alloys is ∿ 40%

For imperfect combustion, these figures are lower.

TABLE 4.7 PERCENTAGE AVAILABLE HEAT WITH VARIOUS FLUE GAS-TEMPERATURES

Flue gas temperature		% fuel available as useful heat	
°F	°C	Perfect combustion	20% excess air
200	93	86	86
400	204	83	82
600	315	79	77
800	427	74	72
1000	538	70	67
1200	649	66	62
1400	760	61	56
1600	871	56	50
1800	982	52	45
2000	1093	46	38
2200	1204	41	33
2400	1315	36	37

Control of Stack Gas Temperature and Composition

Stack gas temperature should be as low as possible, consistent with melting in a reasonable time. Temperature is controlled by the rate which fuel is burned, which is turn depends upon the burner size. The burner should, therefore, be of a size appropriate to the furnace to obtain maximum efficiency. Similarly, the correct fuel/air mixture should be used which can be controlled by stack gas analysis measurement.

Without stack gas analysis equipment, control of furnaces to give maximum efficiency can be difficult, because the methods of controlling the air and fuel supply on some furnaces can frequently be inexact. Figure 4.12 shows the calibration of a gas valve on a furnace. The control consists of a lever operating over a quadrant-type scale having five graduations. Figure 4.12 shows the gas rate against the valve setting which is seen to be non-linear.

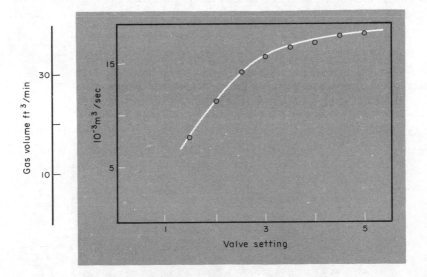

FIG 4.12 300 kg FURNACE CALIBRATION OF GAS VALVE

Figure 4.13 shows the same graph on which is superimposed the stack gas O_2 percentage obtained at the various valve settings.

It will be seen that the optimum fuel efficiency obtained at 0% O_2 is a valve setting of 2, and that air-rich and fuel-rich figures are obtained very quickly at settings other than this value. The furnace was in fact being operated by the furnacemen at a valve setting ~3¾, which used 25 - 30% more fuel than the correct valve setting.

Valves on oil-burners can also be inexact. Some oil-burners do not have a variable oil flow control, but rely on the replacement of a nozzle of standard sizes. It is not unknown for the different size nozzles to become mixed in store and for the incorrect size to be fitted. Figure 4.14 shows the flow rate for different nozzles and how this varies with the oil pressure.

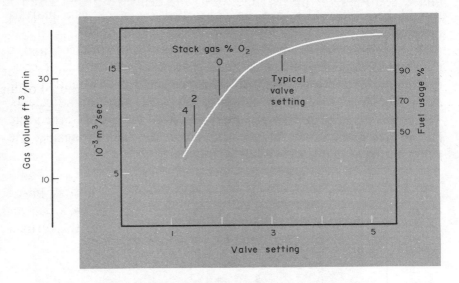

FIG 4.13 300 kg FURNACE CALIBRATION OF GAS VALVE

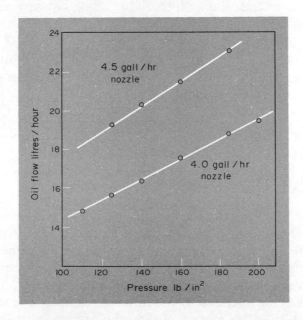

FIG. 4.14 RELATIONSHIP BETWEEN PRESSURE, OILFLOW RATE AND NOZZLE SIZE

Oil pressure is a frequently overlooked important variable and many burners do not have a fitted pressure gauge. The calibration of valve controls, measuring the effect of pressure, etc., is a useful exercise for a technician to undertake and if combined with stack gas analysis readings a good idea of potential fuel savings can be obtained.

The burner on the gas-fired furnace previously mentioned did not have control over air volume. Air was delivered from a fan at whatever volume the fan was producing. Thus the furnaceman could only control the fuel supply, and this even with the optimum fuel/air ratio, could only give a flame of a specific size, no "turn-down" was possible. Burners of this type are virtually uncontrollable, and can be very wasteful of fuel. A burner should have a separate control of air and fuel which can be connected by a linkage system such that air and fuel flow move in unison. Burners can be obtained which are controlled by instruments which measure the volume flow of fuel and air and relay the signals to a fuel/air proportioning instrument.

Another type of burner wasteful of fuel is the type which when controlled by a pyrometer in the molten metal operates the burner "cut-off" by control of the fuel supply only and not both fuel and air. Burners of this type, therefore, push cold air into the furnace, cooling the refractories and hence waste heat and fuel.

Recuperation

The heat of the exhaust stack gases can be used to preheat the incoming supply of air to the furnace through a heat exchanger called a recuperator. If the incoming air is heated to 200 - 400°C, the fuel saving can be 10 - 15% for the typical stack temperatures used in diecasting. The relationship between stack gas temperature savings and air temperature is shown in Fig. 4.15.

Furnace Pressure Control

This is a rather neglected area. Pressures less than atmospheric can develop in furnaces due to the natural buoyancy of the hot gases, and this causes cold air to be drawn into the furnace through charging doors, chance openings, etc., resulting in furnace losses. Pressure can be controlled by a damper in the exhaust stack and should be set to give ~ + 1 mm water gauge. The relationship between furnace pressure and fuel consumption is shown in Fig. 4.16.

Metal Temperature

This should be as low as possible consistent with metal quality. Pyrometers should be used for temperature control.

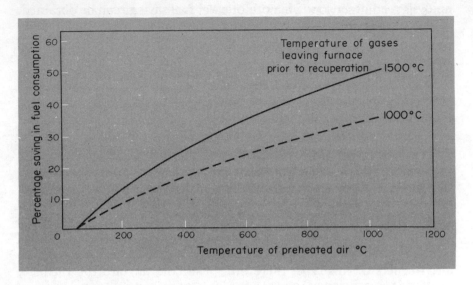

FIG. 4.15 THE SAVING OF FUEL OBTAINED BY PREHEATING OF THE AIR

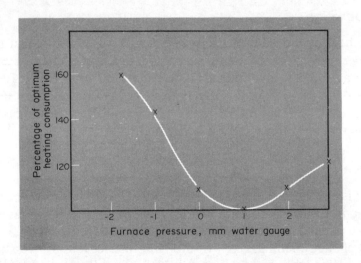

FIG 4.16 EFFECT OF VARIATION IN FURNACE PRESSURE ON FUEL CONSUMPTION

Wall and Metal Surface Losses

Metal surface losses can be substantially reduced by a cover over the crucible top, whilst wall losses can be reduced by modifications to the refractories. If we consider case temperature only, obviously a thicker refractory lining will produce a lower case temperature. However, this will only cause more heat to be stored in the furnace, so thought should be given to the use of refractories of low thermal mass and conductivity which will lower both case temperature and minimise heat storage in the furnace.

Cost of Improvements

Some improvements to furnace efficiency can be obtained for very little expense, and they should naturally be carried out. However, some improvements may be costly, so it is important to balance the cost of fuel against the cost of improvements. This can easily be worked out. Table 4.8 shows the cost of melting aluminium at a fuel price of 30p/therm (1980) against the percentage efficiency of the furnace.

> at   5% efficiency the cost is £66/ton
> at 10% efficiency the cost is £33/ton
> at 15% efficiency the cost is £22/ton
> at 20% efficiency the cost is £16/ton
> at 25% efficiency the cost is £13/ton

TABLE 4.8   MELTING COST IN £ FOR FUEL AT 30p/THERM FOR VARIOUS FURNACE EFFICIENCY AND TONNAGE THROUGHPUT OF ALUMINIUM

Tons/yr	% furnace efficiency				
	5	10	15	20	25
50	3300	1650	1100	800	650
100	6600	3300	2200	1650	1300
200	13,200	6600	4400	3300	2600
400	26,400	13,200	8800	6600	5200
800	52,800	26,400	17,600	13,200	10,400
1600	105,600	52,800	35,200	26,400	20,800

The law of diminishing returns is seen to operate, since to improve the efficiency from 5% to 10% will save £33/ton, but to improve by 5% between 20% and 25% only £3/ton will be saved. A careful balance must therefore be struck between fuel cost, furnace improvement costs and tonnage melted. It is also important to know the existing efficiency, which can act as a base line against which improvement can be measured. It is also important to know whether low efficiency is due to the furnaces themselves or to other foundry variables. If it is due to the latter, it is no use installing more efficient furnaces until these variables have been tacked. The actual saving for various initial furnace efficiencies and tonnage of aluminium/year is given in Table 4.9. Tables such as this should be constructed prior to investing money on furnace improvements, or indeed before buying a more sophisticated furnace which claims to give improved efficiency. The tables can easily be adjusted as the price of fuel is increased, and when this occurs it will readily be seen that improvements not considered worthwhile at a particular fuel cost become financially attractive as this cost is increased.

TABLE 4.9 TOTAL CASH SAVING IN £ BY ACHIEVING 25% EFFICIENCY WITH
FUEL AT 30p/THERM WHEN MELTING ALUMINIUM

Tons/yr	Original % furnace efficiency			
	5	10	15	20
50	2650	1000	450	150
100	5300	2000	900	350
200	10,600	4000	1800	700
400	21,200	8000	3600	1400
800	42,400	16,000	7200	2800
1600	84,800	32,000	14,400	5600

REFERENCES

1. Morgan melting furnaces for the aluminium diecaster.
2. Morgan melting furnaces for zinc alloy diecasting.
3. Energy conservation in fuel-fired furnaces in the non-ferrous metals industry. *Metallurgia & Metal Foundry,* Oct. 1976, pp. 352-353.
4. R. H. Essenheigh, Optimum efficiencies of furnaces: setting the targets. *Industrial Heating,* Nov. 1974, pp. 21-24.
5. R. L. Bennett, Conserving fuel by proper application of combustion equipment, *Trans. SDCE,* 1975, Paper GT 75-041.

Chapter 5

Health and Safety in Pressure Diecasting Foundries

This is a very wide ranging topic and its importance has in the United Kingdom at least, grown considerably in recent years. It is at the same time somewhat ill-defined and yet all-embracing, since under the general heading of Health and Safety come such diverse activities as painted lines on the factory floor, eating at the place of work, protective clothing, the guarding of machines, the disposal of effluent, the discharge of fumes exposure of workers to noise, etc.

Various aspects are covered by national legislation or by local authority by-laws. There are numerous Codes of Practice[1]* which cover various specialised areas and in the United Kingdom there are H.M. Factory Inspectors whose job is to persuade companies to conform to the statutory requirements, but if persuasion fails they do have powers of enforcement. Outside the factory fence the local authority is responsible and it has the powers to enquire regarding effluent disposal and the discharges from chimney stacks, etc.

Generally speaking, a technician will not be directly responsible for health and safety (except that as individuals we all are responsible), nor will the individual be involved with the less quantifiable aspects, but may be called upon to make measurements in certain areas to see if statutory requirements are being met. An example of this could be in the measurement of noise levels. Similarly, it may be necessary to check equipment used for health and safety purposes to find out if it is doing its job. A typical example would be the measurement of air speeds in fume extraction systems over diecasting machines or furnaces. In addition, a technician must be familiar with the safety procedures on any equipment with which he is working. The very nature of the job often means that his is working with equipment when some

---

*Superscript numbers refer to References at end of chapter.

of the safety systems have been removed or overridden, and it is important to realise how such features affect safety.

In the United Kingdom the Health and Safety at Work Act 1974 is a legislative framework aimed at promoting, stimulating and ensuring a high standard of health and safety at work. The Act is an enabling measure, superimposed over existing health and safety legislation. Some features of the Act of interest to the diecasting industry are as follows:

1. Employers must provide training and information and written statements of their policy on health and safety.
2. The general public is protected against harm from work activities, including harm from emissions.
3. Designers, manufacturers, importers and suppliers of articles or substances for use at work must ensure so far as is reasonably practicable that they are safe. They must also supply information about their uses, and this also applies to anyone who builds or installs an article.
4. Employees must take reasonable care for the health and safety of themselves and other persons and to co-operate with employers to enable their duties to be performed or complied with. They must not interfere with or misuse anything provided in the interest of health and safety.

Health and safety in diecasting can be considered to fall into two broad categories. The first cover the more obvious hazards related to guarding of machines, transfer of molten metal, etc., whilst in the second category are environmental aspects such as control of noise or fumes, etc. The more obvious precautions have been well documented,[1] but there is often little information available on environmental aspects.

1. SAFETY ON DIECASTING MACHINES

There are four hazards involved in the use of diecasting machines.

1. Molten metal ejection during injection caused by "flashing" of the die.
2. Molten metal ejection during opening of the die if freezing of the slug is incomplete.
3. Danger of entrapment of the operators person between the platens.
4. Danger from other moving parts, such as the toggle system.

In practice, the toggle system is easily shielded by fixed guards, whilst movable guards are used to prevent metal splashing. The movable guards should not constitute a hazard, and they frequently have a safety trip bar at the leading edge to open the guard again if obstructed on closing. The guards are so arranged that until they are in the fully closed position the die-locking piston will not move. This can be achieved by a mechanical interlock between the guard and a dump valve on the pressure side of the locking system. The valve is open and hence no pressure can be developed to move the locking piston until the guard is at the fully closed position and the dump valve is closed. Some machines also incorporate electrical interlocking operated by long travel limit switches which de-energise the platen closed solenoid in all but the fully closed position.

Systems such as the above should safeguard against hands or tools being trapped between the platens as well as guarding against metal splashing but as an additional safeguard a mechanical scotch or platen arrester is also incorporated. The scotch is progressive in its action so that it is operational even at small die openings.

A typical pneumatic circuit for a diecasting machine guard system is shown in Fig. 5.1.[1] When working with diecasting machines it is

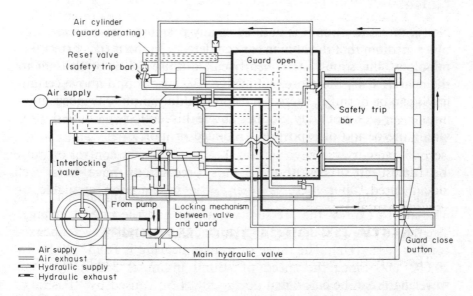

FIG. 5.1 A FULLY SEQUENTIAL PRESSURE/EXHAUST INTERLOCKING CIRCUIT FOR A
GUARD WITH SAFETY TRIP BAR

best to be no nearer than is necessry, and this is why the diecasting U.V. instrument system has leads 15 - 20 ft in length. For preference, it is best to stand at the locking end, and hence well away from the injection plunger and metal transfer area. (See fig. 3.1)

Common sense and an eye for safety are the best guides. It should always be remembered that the injection system can eject metal faster than the operator can run away from it and also that, with many tons force on the locking cylinder, anywhere between the platens can be decidedly dangerous.

The use of metal transfer systems and robots can complicate the guarding situation, as can reciprocating die spray equipment. If a robot is operating from the front of the machine, the standard moving guard is retained, but a guard has to be installed around the robot. If the robot is at the rear of the machine, the standard fixed guard will need to be replaced by a moving guard to allow the robot access. The robot will need to be guarded as if at the front of the machine. Frequently a robot will operate in conjunction with a trim press and the entire area is guarded, the guard at the rear of the diecasting machine frequently being eliminated.

2. ENVIRONMENTAL ASPECTS

Noise

Noise, or rather sound, is defined as any pressure variation in air or other medium that the human ear can detect. Sound is the correct terminology, but sound is frequently called noise if it is unpleasant to the hearer. It is, therefore, subjective. Because sound above certain levels can be harmful, there is a recommendation of 90 decibels maximum over an 8-hour day. Sounds above this level are permitted, providing the period of exposure is reduced proportionally.

Measurement of Sound

Since sound is a variation in pressure on the human ear, it is necesary to express the pressure variations and the rate of change. The rate of change is defined as frequency, measured in cycles per second or hertz (Hz). The range of human hearing is from ~ 20 Hz to 20,000 Hz. Since the speed of sound in air is 340 m/sec, the wavelength can be calculated.

$$\text{Wavelength} = \frac{\text{Speed}}{\text{Frequency}}$$

At 20 Hz the wavelength $= \frac{340}{20} = 17$ m.

Pressure is measured in micropascals (μPa) and the human ear can detect down to 20 μPa, which is a factor of 5,000,000,000 less than the normal atmospheric pressure of 1 kg/cm^2. At the maximum level, the ear can tolerate levels in excess of 20,000,000 μPa. From a practical point of view, to measure sound in μPa would lead to large and unmanageable numbers, and to avoid this the decibel (dB) scale has been devised. The decibel scale uses the 20 μPa threshold as the starting zero-point of 0 dB and each time the sound pressure is multiplied by 10 add 20 dB to the dB scale. The two scales are shown in Fig. 5.2[2]

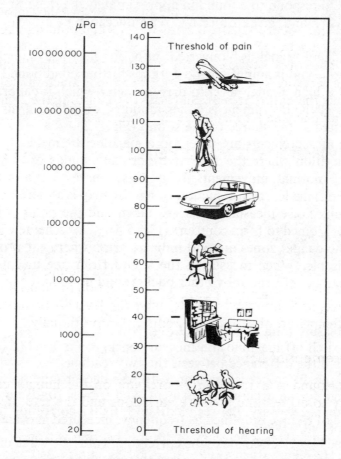

FIG. 5.2 HEARING SCALES

How Sound is Heard

The factors which determine the subjective loudness of a sound are complex. The human ear is not equally sensitive to all frequencies, but is most sensitive in the 2 kHz to 5 kHz range and less sensitive at both the higher and lower ranges. These is also the question of "apparent loudness", which is a relationship between dB and frequency. To give the same apparent loudness as a 1000 Hz frequency at 70 dB, a 50 Hz frequency must be 15 dB higher.

Another factor is sound duration. If less than 1 sec, it is termed impulse sound and the perceived loudness decreases with short impulses of below ~ 70 msec.

Measurement of Sound

This is achieved by a sound level meter which is an instrument designed to respond to a sound in approximately the same way as the human ear, and gives objective reproduceable measurements of sound level.

The sound signal is converted into an electrical signal by a microphone and is amplified and perhaps further conditioned before being of a high enough level to drive a meter which is calibrated in dB. In use, the instrument is either mounted or held at arms length and pointed at the source of the sound (Fig. 5.3).

When measurements are taken on a machine the meter should be at the position where the operative normally stands or within the range of normal movement. If a noise survey of a complete workshop is made, the production of a noise map is a useful exercise. Here many noise measurements are taken and the points of equal noise level joined to form contours (Fig. 5.4).[2] This scheme will show where the danger zones are and indicate areas where ear protectors are advisable. When measuring the sound from one machine, the results can be influenced by other background noise in the area, and in these cases a correction has to be applied as follows:

1. Measure total noise level $(L_S + N)$.
2. Switch off machine and measure background level (L_N),
3. Find the difference between the two readings.
4. If the difference is less than 3 dB, the background noise is too high for an accurate assessment, and if over 10 dB no correction is necessary. Between 3 dB and 10 dB a correction factor is applied.
5. See Fig. 5.5.[2] To make the correction, enter the value obtained from 3, i.e. $(L_S + N - L_N)$, on the botton of the graph. Go up vertically to the line and then horizontally to give the value to be substracted from the total noise level to give the noise level of the machine.

For example, Total noise = 60 dB
Background = 53 dB
Difference =   7 dB

Therefore correction from chart at 7 dB = 1 dB.
Therefore noise of machine = 60 - 1 dB = 59 dB.

The addition of sound can also be a problem. If the noise of two machines is known, it is possible to calculate their joint effect. This is

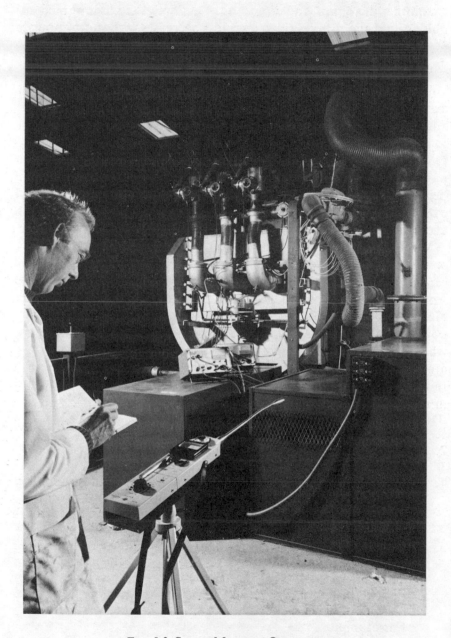

FIG. 5.3 SOUND METER IN OPERATION

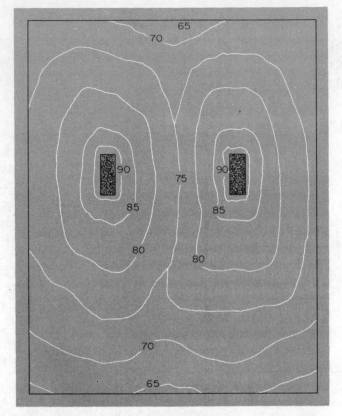

Fig. 5.4 Typical Noise Map Measuring Noise from two Sources

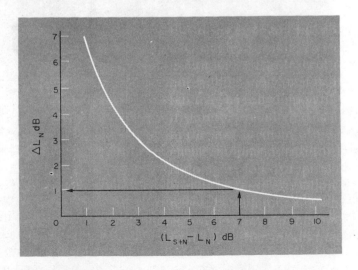

Fig. 5.5 Correction Factor for Subtraction of Noise

useful if the existing workplace is near the 90 dB limit and further plant is to be installed.

The procedures is as follows:

1. Measure levels of individual machines.
2. Obtain the difference in levels.
3. See Fig. 5.6. Enter the difference obtained on the bottom of the graph. Go up vertically to the line and then horizontally to obtain another value, L dB.
4. Add this value to that obtained from the noisiest machine to give the sum of two machines.

   For example, Machine 1 = 85 dB

   Machine 2 = 82 dB

   Difference = 3 dB

   Therefore L dB from graph = 1.7 dB

   Total noise = 85 dB + 1.7 dB = 86.7 dB.

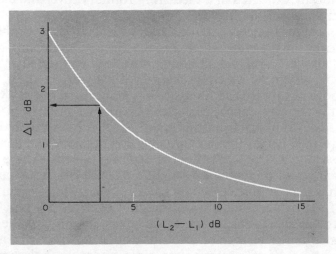

FIG. 5.6 CORRECTION FACTOR FOR ADDITION OF NOISE

NOISE IN DIECASTING FOUNDRIES

There is little published information on noise levels in diecasting foundries, but in two foundries[3] investigated, typical values were obtained as follows:

General diecasting machine noise	= 82 - 88 dB
Air exhaust from pneumatic diecasting machines	up to 93 dB
Die spraying equipment	up to 92 dB
Air fans on burners	= 80 - 90 dB
Trim presses	= 84 - 91 dB
Air operated drills	= 82 - 89 dB
Vibratory finishing	= 89 - 96 dB

If this is representative of most diecasting foundries the situation with the 90 dB limit appears reasonable, since the higher levels are of only short duration. However, in some countries there is a lower limit of 85 dB, and if this limit should become the international limit, diecasting foundries would probably have to improve.

The higher levels associated with the use of compressed air are worth noting as is the noise due to the exhaust on pneumatic machines. The noise of an air jet varies as the 6th - 8th power of air velocity. The noise level in a vibratory finishing shop is expected, since it is an inherently noisy process, and obviously operatives should wear ear protectors.

FUME EXTRACTION

Apart from general overall ventilation there are two areas where fume is locally extracted in a diecasting foundry: over the diecasting machines and over the furnaces.

The machines frequently have a hood suspended above the upper tie bars over the die area, which also acts as a guard against metal if the die "flashes". Individual hoods are often connected to a common trunk system by which discharge is finally effected.

A hood will capture the fume arising from the die spray which is driven upwards by convection currents due to heat from the die block. If the trunking from the hood were without restriction and vertical to roof height (i.e. a chimney), it is possible that convection currents would do the job without assistance, but since there are frequent changes of direction in the trunking, velocity is lost and convection has to be supplemented by a fan system. A fan system is mandatory if the capture hood is not directly over the die block but is located to one side to facilitate die movement. In this case the fan has to give a complete change of direction to the upward current of air and hence must be of greater power, since under these circumstances much uncontaminated air will be drawn into the system. There is no standard way to design a hood system, but the air velocity can be readily measured by a suitable meter to see if it is up to the level stipulated by the designer. The linear air velocity should be measured at the entry to the hood, and this converted to volume in ft^3/min (cfm) by multiplying by the face area of the hood in square feet. If there is a butterfly valve in the system this can be adjusted to give the flow rate needed to extract all the fume. Whether the fume is extracted or not can be established by observation and by the use of smoke testing.

Designing a system is reasonably straightforward, provided it is done in the correct sequence of linear air speed, air volume, fan size, motor size and not by any other method. Figure 5.7 shows a typical system for four diecasting machines. If all have hoods 4ft × 4 ft and hence an area of 16 ft^2, and if to achieve complete fume removal we need an air speed of 100 linear ft/mins, we have the follows situation.

1. Hood no. 1 will pass to duct A, 16 × 100 = 1600 cfm.
2. Hood no. 2 will pass to duct B, 1600 cfm which will also receive 1600 cfm from duct A. Thus 3200 cfm will be passing through duct B and hence to keep a constant air speed will need to be twice the area of duct A.
3. This is repeated for hoods 3 and 4 and ducts C and D.
4. Duct D will need to pass to the fan 6400 cfm, and a fan and motor will need to be chosen based on this figure, due allowances being made for losses in the system due to bends, etc.
5. To achieve the same air speeds at all hoods, in practice a damper or a butterfly valve can be fitted into each duct and its position adjusted based on air velocity measurements with a meter.

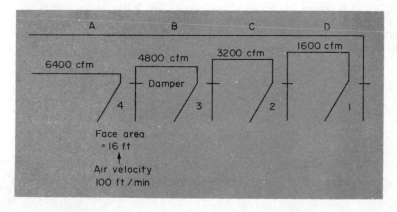

FIG. 5.7 TYPICAL EXTRACTION SYSTEM

The extraction of the fume above furnaces also follows the same principles, but here there is a greater thermal effect from the furnace. The two chief pitfalls in furnace hood design are to make the hood area small, and not to make sufficient allowance for the volume of combustion products produced by the burner. The size of the hood should be such as to allow adequate overlap without drawing in an excessive amount of clean air. The question of insufficient allowance for combustion products is probably the most common fault, and this is indicated by furnace hoods which are full of combustion products and spilling out from the sides.

A furnace using 20 therms of natural gas/hr has ~ 100,000 ft^3 of combustion products produced per hour at ~ 1000°C, i.e. ~ 1600 cfm, and this must be allowed for in the calculations.

It may be necessary to sample the fumes in the stack to check what fumes are being discharged. Similarly, personal sampling may be needed on operatives who are exposed to substances which have a threshold limit value. In diecastings this is only likely to be furnace operatives who are constantly in contact with fumes arising from fluxes, which frequently contain chlorides and fluorides of the alkali metals.

The philosphy with all aspects of health and safety is that it is better to be safe before than sorry after the event. Many aspects of safety are simply applied common sense, but some are not, and it is in these latter areas where the technician can be of help not necessarily by doing the investigational work himself but by knowing where to go for specialist advice.

REFERENCES

1. *Safety in Pressure Diecasting.* 5th edition. Zinc Alloy Diecasters Association, 34 Berkeley Square, London, W1.
2. *Measuring Sound.* Brüel & Kjaer, Noerum, Denmark.
3. T. B. Hill. Private communication.

Chapter 6

Some Applications of Statistics to Pressure Diecasting Foundry Variables

The application of statistical thinking and techniques can be useful to diecasting foundries, and the full potential has hardly yet been appreciated by much of management. Statistics can be used in the sphere of quality control to assess process variability and to monitor changes in metal composition reject rate or casting dimensions and to detect trends before they become serious. They can be used to set up inspection schemes based on statistical probability, whilst in research and development activities they can be used to plan experiments to get the maximum benefit from the minimum number of results as well as to assess the significance of results when obtained.

As in most specialised areas there is no substitute for a competent statistician when setting up a quality control scheme or when analysing the results of experimentation. Similarly, it is necessary to become familiar with the statistical ideas and terminology by reading the publications on statistics which also contain the necessary tables and graphs not reproduced here for lack of space. However, it is possible with the aid of a calculator for technicians to carry out simple statistical calculations which when allied to common sense and experience can point the way to a particular line of action which may not have been realised without the statistical information.

There are three broad areas in diecasting when the use of statistics can find ready application, these are:

1. The assessment of process variability.
2. Inspection and quality control.
3. The analysis of "on-plant" experimentation.

1. ASSESSMENT OF PROCESS VARIABILITY

Any manufacturing process making any component will not make all components to the same size or shape, which is not to say that rejects are being produced. Engineers and designers understand the above fact of life and take account of it by having a tolerance limit on components, which allows for deviations from the ideal situation up to a given value. The reasons why all components are not the same is due to the inherent variability in the manufacturing process itself, due to such factors as variation in input materials, operator skill, equipment performance, etc. Diecasting is no exception to this rule, and tolerances are frequently imposed on casting dimensions, metal specifications and casting defects. It may be thought surprising that casting defects can be considered as a "tolerance", but it is so since in the ideal case there are no defects and the acceptance of less than perfect casting must represent a "tolerance".

In diecasting, variability can be caused by machine variables such as machine wear, locking and injection force variations, or by die variables such as temperature, lubrication, wear, etc. The sum total of all the random variables gives rise to variability with regard to casting quality and being of a random nature cannot be assigned to a specific cause.

Take as an example a particular dimension on a casting which should be 0.500in. If we measure fifty castings we could achieve the results shown in Table 6.1. [It should be noted that these figures are

TABLE 6.1 NUMBER AND DIMENSIONS OF CASTINGS

Actual size, in	Number of castings
0.485	2
0.490	4
0.495	12
0.500	16
0.505	10
0.510	5
0.515	1

for calculation purposes. Actual diecastings are very much more accurate[1*].] The castings will all be acceptable providing the dimension is specified as 0.500in ± 0.015in. However, it is possible that if the same die were put on a different machine a different size distribution

*Superscript numbers refer to References at end of chapter.

TABLE 6.2 NUMBER AND DIMENSIONS OF CASTINGS
PRODUCED BY THREE MACHINES

Actual size	Number of castings		
	Machine A	Machine B	Machine C
0.475			
0.480			
0.485	2		
0.490	4		
0.495	12	14	2
0.500	16	20	4
0.505	10	16	12
0.519	5		16
0.515	1		10
0.520			5
0.525			1

would be obtained. In three machines, the situation could occur as shown in Table 6.2.

Machine A will produce good castings with a tolerance of ± 0.015in. Machine B will produce good castings with a tolerance of ± 0.005in. Machine C will produce good castings with a tolerance of − 0.005in + 0.025in.

Thus if the tolerance were ± 0.005in, machine A would produce 12 rejects, i.e. 24%, machine B would produce no rejects, and machine C would produce 32 rejects, i.e. 64%. Machines A and C are operating normally in the sense that there is no assignable cause for the high reject rate. It is due to natural variability being greater and so a greater variability in the final produce is produced.

It is important to know whether defective castings are being produced because of natural variability or whether there is an assignable cause; to know this the natural variability has to be assessed.

When Table 6.1 is plotted in the form of a histogram (Fig. 6.1) it is seen that the results approximate to a normal distribution and this is the key to the assessment of variability.

A typical normal distribution curve is shown in Fig. 6.2. The curve has two main properties represented by $\overline{X}$ and σ.

$\overline{X}$ is the mean value of the distribution being dealt with (i.e. the arithmetical average) and σ is the standard deviation. Standard deviation is a measure of the spread of results around the mean. Thus two sets of results can have the same mean value, but if one set has more spread than the other, its standard deviation will be greater. The standard deviation of a set of results is calculated as follows:

1. Calculate the mean of the set of values.
2. Subtract the individual values from the mean value.
3. Square the individual differences.
4. Add together the values from 3 to give the sum of the squares.
5. Divide the sume of the squares by the sample number *n,* or if the sample size is small by *n*-1, to give the sample variance.
6. Find the square root of the sample variance to give the standard deviation.

The above calculation can be time-consuming, particularly if due account is taken of several decimal places which is needed for accuracy. The use of a suitable calculation which will produce $\overline{\times}$, σ_n and σ_{n-1} is therefore recommended. The differences between σ_n and σ_{n-1} will also be readily seen, and the consequences of using either value assessed. If in doubt, however, it is safest to use the value based on σ_{n-1}.

The normal distribution curve has the property that only $\sim 0.2\%$ of results exceed $\pm 3 \sigma$ and, therefore, if a particular dimension has a tolerance of 3σ, only 0.2% rejects will be produced. This is a commonly used value and is sometimes referred to as the inherent capability or natural tolerance of the process. So in a stable process

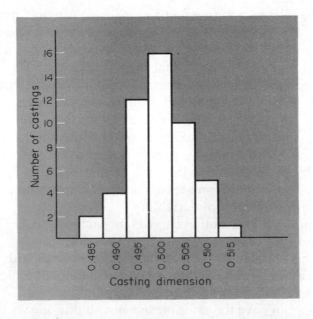

FIG. 6.1 HISTOGRAM PLOT OF RESULTS IN TABLE 6.1

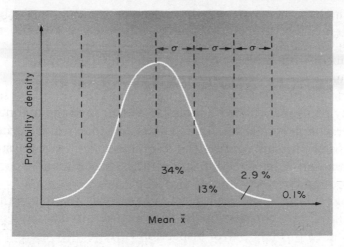

Fig. 6.2 Normal Distribution Curve

(i.e. with only random variables) practically all components will be produced within a tolerance range equal to 6σ.

The importance of establishing the value of 6σ is seen by the following example.

If one particular machine and die combination produces components of $\bar{x} = 0.500$in and $\sigma = 0.001$in, a 6σ band gives components between 0.497in and 0.503in.

If the specification calls for 0.500in $\pm$ 0.005in no rejects will be produced, since the inherent process capability is within the specification tolerances. If, however, $\bar{x} = 0.500$in and $\sigma = 0.003$in, a 6σ band gives components between 0.491in and 0.509in, so if the specification calls for components to be 0.500in $\pm$ 0.005in the reject rate will be high simply because the process variability is outside the specification tolerance. The actual reject rate can be found from published values in statistical tables.

Calculation of the standard deviation is rather tedious where large sample sizes are involved, and this has led to the adoption of short cuts. Two common alternatives are to express the spread of results by using the sample range, i.e. the largest size minus the smallest size, or to calculate the standard deviation by the group frequency table method. This method has been used[2] to calculate the $\pm 3\sigma$ spread on actual diecastings.

A knowledge of the process variability is useful when there are changes in the parameter being measured. If it is known from measurements that the 6σ spread is within the specification, but inspection is showing that castings are beginning to be produced having dimensions outside this limit, it is necessary to look for an assignable cause which is not part of the inherent variability. Without a knowledge of the process variability beforehand, it would not be known if an assignable cause was present or not.

2. INSPECTION AND QUALITY CONTROL

Quality control should be applied at all stages of the manufacturing process, and should be used to assess reasons for rejects on the factory floor. In this context statistics is a useful tool to measure trends in a particular direction so that these trends can be corrected before too many rejects start to occur.

Castings should always be inspected as early as possible in the production line, since this avoids the cost incurred by subsequent processes, and to this end patrol inspection is frequently used. This involves a patrol inspector moving around the diecasting shop examining castings. Sometimes all castings are examined, sometimes a representative sample decided by a prearranged sampling procedure.

With anything less than 100% inspection, the method of sampling, the numbers to be examined, the batch size and the accept or reject levels should always be arrived at by discussion with the customer, since any sampling scheme involves two types of error called by statistitians type 1 and type 2 errors. Type 1 error occurs when the sample for inspection has more proportionally defective castings than the batch from which it is taken. This causes a wrongful rejection of the batch. Type 2 error occurs when the sample for inspection has less proportionally defective castings than the batch from which it is taken. This causes a batch to be released to the customer with which he will be less than happy. The above errors are sometimes referred to as the manufacturer's risk and the consumer's risk. In reality the reject batches are not sent for remelting, they are simply submitted for further sampling or to 100% inspection. Because of the above risks it is important that they are quantified on the basis of long-term probability, so that an acceptable plan can be agreed. Anything less than a sampling scheme based on statistics and taking into account acceptable quality levels (AQL) and quantifiable manufacturer's and consumer's risks is bound to be arbitrary and can lead to disputes between the parties concerned. The advantages of a statistically based inspection scheme are, however, very great. The manufacturer can substantially reduce the amount of inspection carried out and hence the number of inspectors, whilst the user has confidence that the released batches of castings can go immediately onto the production line without having to be inspected, and the batches received will in the long term contain no more than a known percentage of defective castings.

From a consumer's point of view one of the chief problems is having to deal with castings which, have a wide batch-to-batch variation in the number of defective castings, and frequently as a safeguard he may wish to examine all the castings before releasing them to his production line. He will then return all defect castings to the supplier. The consumer's complaint is that he is carrying out the inspection which should have been completed by the supplier.

From the supplier's point of view there is a danger that if the consumer inspects a batch of castings and the first few inspected are rejects, the entire batch may be returned out of shear exasperation. Neither of these situations should occur if a properly worked out and agreed inspection scheme based on statistics is operated.

There are alternative schemes which can be used, depending upon the reasons why castings are rejected, but generally for castings a sampling for attributes is required. This means that a casting is either accepted or rejected. the purchaser is not concerned about the degree of defectiveness, i.e. whether a casting has one defect or several defects, it is rejected. The purchaser does not differentiate between the two.

The supplier, however, should be concerned about the degree of defectiveness, since if castings are being rejected for say cold finish in two areas, if this defect can be cured only in one area the castings will still not be accepted by the customer.

Before a sampling scheme is put into operation the answers to the following questions should be known.

1. The acceptable quality level (AQL) required.
2. The inspection level.
3. The batch size.
4. Whether single, double or any other type of sampling is to be used.

The Acceptable Quality Level

This is defined as the only-just-acceptable quality. It is the poorest quality that is considered acceptable as the process average.

The Inspection Level

This is the relationship between the batch size and sample size and is obtained from standard tables. The tables are such that as the batch size becomes larger, the sample size becomes larger, but not by the same ratio. In standard tables the inspection levels are designated I, II and III and generally level II is used.

The Batch Size

This is open to discussion depending upon many factors. It could be the total castings produced from one machine in 1 hr or in a day, depending upon circumstances. It is, however, not advisable to increase the batch size by combining the output of castings made from say two machines producing the same castings, since if one is producing more defectives than the other the general batch reject rate could be increased.

The Sampling Plan

This is decided in advance. Frequently a single sampling plan is used with a given batch and sample size. However, there are advantages to be had by a double plan. Here the first sample taken is smaller than for a single plan, and if the quantity of the sample is good or bad, the batch is released or rejected. If of intermediate quality, a second sample is taken before a final decision is made.

An Example of a Typical Scheme

There are British Standard specification[3, 4] which detail sampling procedures, the following being a typical case.

We assume there is agreement between supplier and purchaser as follows:

1. AQL level = 1.5% rejects.
2. Inspection level = standard II.
3. Batch size = 600 castings.
4. Single sampling plan.

Having decided upon the above the decision is related to BS 6001 [4] where the appropriate table is table X-J-2 where the following parameters are obtained.

1. The sample size is 80 castings.
2. The batch is accepted if 3 castings or less are defective.
3. The batch is rejected if 4 castings are defective.

Thus for every batch of 600 castings, 80 at random are examined and if 3 or less are defective, the entire batch of 600 castings is passed. If 4 are defective, the batch is rejected, i.e. the entire batch is subjected to 100% inspection.

The above can represent a considerable saving of labour on inspection, since the inspection of 80 castings is no more than 13% of the effort of inspecting 600. For batches where defectives are high, the inspection of 80 castings is unnecessary since inspection only proceeds until 4 defective castings have been found, at which point the entire batch is rejected. Since the worse the batch the earlier the 4 defectives will be found, it means that inspection labour is not wasted examining poor quality batches. In addition to a saving on inspection, the consumer is guaranteed that in the long term the batches received will contain 1.5% rejects, which was the AQL level agreed in the sampling plan.

If the agreement is for a different AQL, the number of reject castings per sample will change, e.g. for an AQL of 4% with a batch size of 600 and a sample size of 80, the batch will be accepted if the sample has 7 or fewer defectives and rejected with 8 defectives.

Any sampling plan such as the above can be described by an operating characteristic curve (Fig. 6.3), which shows the probability of batches accepted for a given percentage defective rate in the batches examined. Figure 6.3 shows the operating characteristic curves for 1.5% and 4.0% AQLs.

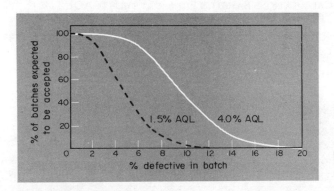

FIG. 6.3 OPERATING CHARACTERISTIC CURVES FOR SINGLE SAMPLING PLANS

On the first examination of the characteristic curve it may seem strange that if say, a 4% AQL level has been agreed, the sampling scheme will allow batches to pass to the consumer with up to 20% defectives. This is, however, the case and this is the consumer's risk. Similarly, there is a producer's risk. This is the probability, however small, that batches of better quality than the 4% AQL will be rejected simply because by chance the sample of 80 castings will contain a greater proportion of defectives than the batch from which they were drawn. However, the curve indicates that for a batch containing, say, 20% defectives there is only a 1% probability that it will be sent to the consumer. Similarly, with 6% defectives in a batch there is an 89% chance, and with 4% defective batches there is ~ 98% chance of being sent to the consumer.

These high defective rates are, however, balanced out in the long term by batches of castings which are supplied with less defectives than the 4% AQL, to give an overall rate of ~ 4% AQL. It is possible to demonstrate the above by calculation.

Table 6.3 represents a hypothetical case of castings being inspected in samples of 80 from batches of 600 to an AQL of 4%. Column 4 shows that the foundry is actually producing casting batches containing between 1.6 and 9.55% defectives. However, column 5 shows the probability of the various batches being passed to the purchaser (these probabilities are obtained from the 4% AQL operating characteristic curve in Fig. 6.3) and column 6 shows the number of batches passed to the purchaser based on these probabilities. Column 7 shows the actual number of castings supplied to the purchaser and column 8 the number of defectives based on the figures in column 4 which would be detected by the purchaser on his production line.

TABLE 6.3 CALCULATIONS BASED ON FIG.6.3 AT THE 4% AQL

Batch size	Number of batches	Total number of castings	% defective in batch	Probability passing batch to purchaser, %	Number of batches sent to purchaser	Number of castings sent to purchaser	Number of defective castings sent to purchaser
600	2	1200	3.72	100	2	1200	~ 45
600	4	2400	2.42	100	4	2400	~ 59
600	3	1800	7.50	75	2	1200	90
600	2	1200	5.06	95	1	600	~ 31
600	1	600	1.60	100	1	600	~ 10
600	6	3600	4.00	100	6	3600	144
600	3	1800	2.10	100	3	1800	~ 38
600	2	1200	3.50	100	2	1200	42
600	2	1200	9.55	50	1	600	~ 58

Adding up the totals in columns 7 and 8 will indicate 517 defective castings in the 13,200 castings supplied. Thus the percentage defectives found by the purchaser

$$= \frac{517}{13,200} \times 100 = 3.9\%$$

which is within the 4% AQL value the particular sampling plan is designed to give.

From a supplier's point of view such a plan will produce large savings of inspection labour. Column 2 shows that 25 batches will be inspected of 80 castings/batch, i.e. 2000 castings in all compared to 15,000 castings by 100% inspection. This is a saving of ~ 87%.

If the 100% inspection necessary on the reject batches is taken into account it is necessary to inspect 3 batches completely, i.e. 1800 casting, therefore overall the above scheme would require 2000 + 1800 castings = 3800 castings to be inspected instead of 15,000 castings by 100% inspection, which represents a saving of $\sim$ 75%.

The saving in inspection labour is a strong incentive for the supplier to keep the quality high, particularly as reject batches must be 100% inspected. From the purchaser's point of view castings supplied by such a scheme are guaranteed over the long term to contain no more defectives than the percentage of AQL, and it is up to the supplier and purchaser to liaise as to what the percentage of AQL should be, taking into account normal commercial considerations.

Both parties must, however, adhere rigidly to the scheme, even if from time to time the user has a relatively high number of defective castings in the batches supplied and is tempted to send them back. Similarly, the supplier must avoid the temptation to pass a batch if the number of rejects in the sample is only one more than the acceptance level. The purchaser can, however, establish if this is being done by the supplier simply by keeping a record of the number of defectives in each batch which he detects. If in the long term the number of defectives is greater than the agreed AQL level, the supplier is "bending" the rules. In effect both parties must agree on a scheme and stick to it in the knowledge that in the long term it is mathematically correct and will benefit both parties. This is not to say that a concession procedure should not be operated if both parties agree. It is well known that a purchaser is less rigorous in his demands if he needs the castings that day!

Irrespective of the type of inspection in operation, a record of the numbers rejected or the percentage rejected should be kept and tabulated or plotted in sequence to determine long-term trends for use in quality control.

In its simplest form the number of castings rejected or the percentage rejected can be plotted on a P chart[2] which shows the reject rate against a time interval. A slightly more elaborate chart would also plot the sample range.[5] Both charts are intended to show trends and are intended to be used as an alarm system to indicate that if the trend continues rejects will result, and this is achieved by lines on the chart which show warning and action limits. The purpose of any chart is to give more visual impact than a series of numbers, and in this regard some types of chart have more impact than others. Take as an example a foundry producing castings with a typical scrap rate of $\sim$ 14%. The actual scrap rate can be plotted in sequence as in Fig. 6.4, in which will be seen the typical day-to-day variations to be expected, but no real obvious trends either upwards or downwards. If, however, a cusum[6] chart (Fig. 6.5) is plotted, the same data will indicate a definite change to higher scrap rates. The plotting of a cusum chart is not as straightforward as a conventional chart, and it

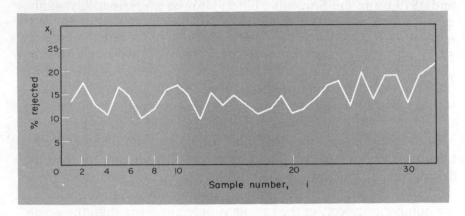

FIG. 6.4 CONVENTIONAL CHART FOR DATA

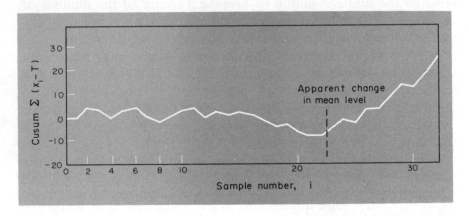

FIG. 6.5 CUSUM CHART FOR DATA

represents a change in the mean level of a particular parameter (in this case it is scrap rate) by a change in the slope. The decision rules as to whether action should be take is based on the magnitude of the change in the slope and the length over which the apparent change has persisted.

This method used to calculate and set up quality control charts can be used to produce any type of chart to assess a trend, and indicate that action should be taken. Figure 6.6 is a cusum chart related to the weight of ingot purchased by an aluminium foundry over a 2-year period.

The chart arose due to the following situation.

The foundry was concerned about its increased oil usage in the melting/bale-out furnaces and wished to know if the furnaces could be improved. A comparison of company records for the years 1978 and 1979 using the Students 'T' test showed no statistically significant differences in the quantity of oil purchased (the cost had naturally increased), but they did show significant differences at the 5% level in the weight of metal ingot purchased, and hence the weight of saleable castings. The foundry was a classic case of working

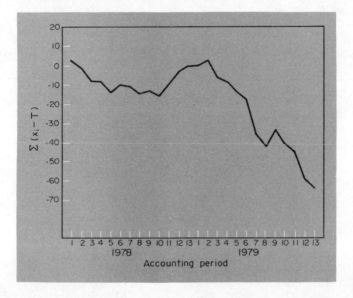

FIG. 6.6 WEIGHT OF ALUMINIUM ALLOY INGOT PURCHASED

the furnaces in the same way irrespective of the metal being demanded by the foundry. The question to answer was not, how could the furnace efficiency be improved, but rather at what point (related to ingot tonnage purchased) did the foundry have to consider working its furnaces in a different way. To assess this point it was necessary to construct a chart to show a trend in ingot purchases/accounting period, since the weight of metal melted/day was unobtainable. Table 6.4 shows how a cusum chart related to ingot purchases was calculated and Fig. 6.6 is the cusum chart.

The chart shows that there is a cycle withing the year consisting of a fall in purchases during the spring, a levelling off in the summer and an increase in the autumn. In 1979 the levelling off in the summer did not occur, and by period no. 7 the foundry has purchased $\sim$ 37 tons less metal than the target value. This was substantially less than at any time in the previous year, and if the chart had been in operation at that time action could have been taken. Since there was no chart to indicate a trend, nothing was done until about period 10 (i.e. 12 weeks later). The chart was based on records already in existence at the foundry and is used to illustrate what can be done. The chart should really be on daily or weekly figures of metal melted to enable action to be taken immediately.

TABLE 6.4 CUSUM CHART CALCULATIONS

Accounting period	x_1	$x_1 - T$	$\Sigma(x_1 - T)$
1 1978	20.2	2.3	2.3
2	14.3	-3.6	-1.3
3	10.2	-7.7	-9.0
4	18.7	0.8	-8.2
5	12.6	-5.3	-13.5
6	21.5	3.6	-9.9
7	17.5	-0.4	-10.3
8	13.7	-4.2	-14.5
9	19.6	1.7	-12.8
10	15.4	-2.5	-15.3
11	23.3	5.4	-9.9
12	24.8	6.9	-3.0
13	21.0	3.1	0.1
1 1979	18.0	0.1	0.2
2	20.6	2.7	2.9
3	9.1	-8.8	-5.9
4	15.6	-2.3	-8.2
5	12.9	-5.0	-13.2
6	13.6	-4.3	-17.5
7	0.1	-17.8	-35.3
8	11.2	-6.7	-42.0
9	26.2	8.3	-33.7
10	10.4	-7.5	-41.2
11	13.9	-4.0	-45.2
12	3.7	-14.2	-59.4
13	14.1	-3.8	-63.2

x_1 = weight of metal purchased in tons
T = mean value in tons/period for 1978 (17.9 tons)

Calculations

1. Arrive at a target figure (T). In this case 17.9 tons, which is the mean value for 1978 which was considered a satisfactory year.
2. Subtract the T value from x_1 values.
3. Add all x_1 - T values taking into account algebraic sign.
4. Plot in sequence against accounting period.

3. ON-PLANT EXPERIMENTATION

The potential for statistical analysis of the variables in a diecasting plant is large, simply because we are dealing with large numbers of castings, machines, furnaces and other variables. There is unfortunately only a moderate amount of statistical thinking applied to everyday problems. There are no standard methods of approach to on-plant experimentation or in analysis of the results as in inspection or quality control, so each case has to be considered individually. However, an example of the type of problem presented and its solution can be obtained by a study of case histories.

The following case histories are examples of the use of statistical thinking and are offered to stimulate rather than to be followed in precisely the same way. No two problems or foundries are exactly alike, so each situation should be thought out as it occurs. Details of some of the tests described and how they are worked out must as before be obtained from publications on statistics, since space precludes their inclusion.

Case History 1 — The Estimation of Furnace Capacity

Background

A company was replanning its zinc diecasting foundry and wished to install a bulk melting furnace.

Question

What size of furnace should be installed?

Answer

Most companies with this problem would simply say "What is my maximum throughput of zinc per day?" They would arrive at a figure and then add on a further amount to allow for error and then purchase the next largest furnace size. This traditional method has in the past generally resulted in furnace capacity which is too large, with a consequential waste of fuel. A statistical approach was, therefore, used to estimate furnace size. The company had eighteen diecasting machines and the weight of metal cast by each was obtained over a 12-week period. Since the machines were of different types capable of varying shot rates and weights, the weights were averaged. The results are summarised in Table 6.5.

TABLE 6.5 METAL REQUIRED OVER A 12-WEEK PERIOD

Week	Metal weight kg/week	Estimated weight kg/day on 5 days
1	25,490	5098
2	22,495	4499
3	21,829	4366
4	25,091	5018
5	25,864	5173
6	23,215	4643
7	23,233	4647
8	22,911	4582
9	26,721	5344
10	24,329	4866
11	20,369	4074
12	23,943	4789

Calculating the mean $\overline{x}$ we have 23,790.
Calculating the standard deviation σ we have 1819.
Therefore 6σ limit = 18,333 - 29,247 kg,
i.e. 3667 - 5849 kg/day for a 5-day week.

Thus it is seen that a furnace of $\sim$ 6000 kg/day would meet all requirements, which is a figure probably not far removed than if the size had been worked out by traditional rule-of-thumb methods, i.e. maximum metal needed (5344 kg) + $\sim$ 10% for error = 5900 kg.

However, the traditional method of calculating furnace size would not indicate that a 6000 kg/day requirement is most efficiently met by have two furnaces.

It can be calculated that with two furnaces, for minimum excess melting capacity the size of the large furnace should be $\sim$ the mean of the sample distribution $\overline{x}$, with the size of the small furnace $\sim 3\sigma$.

That is, large furnace = 4758 kg/day
small furnace = 1090 kg/day

Translating this into "real" furnaces we make little error at 5000 kg/day and 1000 kg/day furnace sizes. The reduction in excess melting capacity using two furnaces is readily demonstrated.

If we use the figures in the kg/day column from Table 6.5 and assume a single 6000 kg/day furnace, we have the situation as shown in Table 6.6.

If, however, we have two furnaces, one at 5000 kg/day and one at 1000 kg/day, we have a different situation, as shown in Table 6.7.

TABLE 6.6. EXCESS MELTING CAPACITY WITH ONE FURNACE

kg/day	Excess metal melting capacity, kg
5098	902
4499	1501
4366	1634
5018	982
5173	827
4643	1357
4647	1353
4582	1418
5344	656
4866	1134
4074	1926
4789	1211
Total 57,099	14,901

$$\frac{\% \text{ excess}}{\text{melting capacity}} = \frac{14,901}{57,099} \times 100 = \sim 26\%.$$

TABLE 6.7 EXCESS MELTING CAPACITY WITH TWO FURNACES

kg/day	Furnaces needed	Excess metal melted capacity, kg
5098	5000 + 1000	902
4499	5000	501
4366	5000	634
5018	5000 + 1000	982
5173	5000 + 1000	827
4643	5000	357
4647	5000	353
4582	5000	418
5344	5000 + 1000	656
4866	5000	134
4074	5000	926
4789	5000	211
Total 57,099		6901

$$\frac{\% \text{ excess}}{\text{melting capacity}} = \frac{6901}{57,099} \times 100 = \sim 12\%$$

Thus the use of two furnaces is seen to lower the excess melting capacity and give a saving in fuel. In actual fact if the foundry were to "plan" the use of the two furnaces, they would avoid using the 1000 kg furnace on two of the four days and they would lose a minimum of production. The excess melting capacity would then be reduced to ~ 9%.

Since the foundry was diecasting zinc, it was possible to eliminate the small furnace completely and charge some zinc ingot at the machines when the 5000 kg furnace was at a maximum, and under these conditions the average excess capacity was ~ 6%.

The above calculations are naturally only the first stages in obtaining information on which to base decisions. Other factors to take into account are the actual furnace sizes available and their efficiency of working under all conditions of loading.

Clearly the above approach is only applicable if the metal demand is described by a normal or nearly normal distribution curve, and whether this is so in each case can be seen if a histogram plot is made of metal demand/day. If the distribution is markedly skew, the use of the mean figure can give a misleading conclusion.

Case History 2 — Impregnation of Castings

Background

A company was purchasing the same type of diecasting from two sources. The castings were extensively machined and had to withstand a pressure test. One supplier was impregnating all castings, whilst the other was not.

Question

Where the impregnated castings producing fewer rejects than unimpregnated castings after machining?

Answer

On the face of it the question should answer itself, since common sense would indicate that impregnation is bound to give an improvement. This is, not always the case, since for impregnation to be effective, the porosity in the casting must appear at the surface and not just be discovered after machining, also the impregnation must be properly carried out. Examination of company records over 32 weekly periods showed that for impregnated castings the average reject rate was ~ 5.9%, with a spread from 1.2% to 39.5%, whilst for non-impregnated castings the average was ~ 8.7% with a spread from 0% to 19.78%.

The figures are given in Table 6.8 and in fact they tell very little except that there is considerable scatter in the reject rate and the values are not necessarily normally distributed. Because of this the Mann-Whitney U Test was used to test for significance between the two sets of results, since this is a ranking test and does not depend upon a normal distribution.

This test is carried out as follows:

1. Both sets of results are combined into one column, in order of increasing reject percent, i.e. 0% is the first and 39.50% is the last.

2. Each percentage reject is given a rank number from 1 at the lowest to 61 at the highest percent. If two or more values are the same, the values are averaged, e.g. in the case quoted there are four instances of rejects at 2.69% and these occur after rank number 14. Since they are of the same value, they cannot be ranked 15, 16, 17, 18. The rank numbers are, therefore, averaged and all 2.69% reject rates are given rank 16½. The next number after 2.69% is given rank 19.

3. Table 6.8 shows the rank numbers given for each reject rate.

4. The rank totals for impregnated and non-impregnated castings are then calculated.

5. For impregnated castings the rank total is 777. For non-impregnated castings the rank total is 1114.

6. Since we have more than 20 numbers in each case we use the following formula to calculate the value Z.

$$Z = \frac{U - \dfrac{n_1 n_2}{2}}{\sqrt{\dfrac{(n_1)(n_2)(n_1 + n_2 + 1)}{12}}}$$

where n_1 = number of samples in smallest group, i.e. 30
n_2 = number of samples in largest group, i.e. 31

$$U_1 = n_1 n_2 + \frac{n_1(n_1+1)}{2} - R_1$$

where R_1 = rank total of smallest group, i.e. 1114

and $U_2 = n_1 n_2 + \dfrac{n_2(n_2+1)}{2} - R_2$

where R_2 = rank total of largest group, i.e. 777.

7. We, therefore, have to calculate the U_1 and U_2 values.

$$U_1 = (30 \times 31) + \frac{30\,(30+1)}{2} - 1114$$

$$U_1 = 930 + 465 - 1114$$

$$U_1 = 281$$

$$U_2 = (30 \times 31) + \frac{31(31+1)}{2} - 777.$$

$$= 930 + 496 - 777$$

$$U_2 = 649.$$

8. We then check that $U_1 + U_2 = n_1 \times n_2$

   i.e. $281 + 649 = 30 \times 31$

   $\qquad = 930 \qquad = 930.$

9. Substitute U values in previous formula and check using U_1 or U_2

   for U_1, $Z = U_1 - \dfrac{n_1 n_2}{2} \div \sqrt{\dfrac{(n_1)\,(n_2)\,(n_1+n_2+1)}{12}}$

   $$Z = 281 - \frac{930}{2} \div \sqrt{\frac{930 \times 62}{12}}$$

   $$Z = -2.66$$

   for U_2, $Z = U_2 - \dfrac{n_1 n_2}{2} \div \sqrt{\dfrac{(n_1)\,(n_2)\,(n_1+n_2+1)}{12}}$

   $$Z = 2.66$$

   i.e. the two values of Z are the same, but the sign is different, depending upon whether we use U_1 or U_2 in the calculation.

10. We then look up in statistical tables the probability associated with $Z = 2.66$.

11. We find the probability to be 0.0039. Therefore the differences are significant at the 0.5% level, but not at the 0.1% level. This, however, is significant enough to justify a recommendation that impregnation be continued and extended to the other foundry.

TABLE 6.8 % REJECTED FOR IMPREGNATED AND NON-IMPREGNATED CASTINGS

Week	% Rejects			
no.	Foundry A (impregnatd)	Rank	Foundry B (not impregnated)	Rank
20	3.22	27	19.78	59
19	2.75	19½	18.52	57
18	2.96	24	18.37	55
17	2.69	16½	9.59	46
16	6.57	38	2.89	21½
15	2.94	23	5.04	33
14	9.04	44	7.29	42
13	6.20	35½	14.14	52
12	6.61	39	6.20	35½
11	1.34	5	2.69	16½
10	—		2.47	12
9	2.89	21½	—	
8	1.26	4	3.63	30
7	1.79	8½	3.44	28
6	1.20	3	9.82	47
5	1.76	7	3.55	29
4	2.69	16½	7.03	41
3	3.82	32	8.37	43
2	2.69	16½	12.63	50
1	2.16	11	—	
51	1.58	6	5.34	34
50	1.79	8½	2.75	19½
49	2.68	14	3.08	25
48	3.69	31	0.00	1½
47	9.24	45	3.14	26
46	2.48	13	6.25	37
45	1.94	10	17.44	54
44	39.50	61	18.40	56
43	6.93	40	0.00	1½
42	12.12	49	16.60	53
41	11.94	48	13.17	51
40	23.30	60	19.24	58
	Rank total	777		1114

Case History 3 — Estimation of Die Temperature at Metal Injection

Background

Castings were being produced having a high proportion of "cold" defects on the surface. Two identical dies on different machines were involved and the impression was that the reject rates were higher on the "new" die as compared to the "old" die. Isolated temperature measurements had shown little difference between the two dies.

Question

Was there a difference in die temperature at metal injection?

Answer

The temperature of a die cavity cannot be measured except by mounting thermocouples in the die block at the die/metal interface and this is difficult on production dies. The temperature, therefore, has to be assessed indirectly, in this case by measuring the casting temperature immediately after ejection. Since the dies and castings weights were identical, the casting temperature at ejection is a reflection of the temperature of the cavity at injection. Fifty temperatures were measured on the "old" die and twenty-five on the "new" die. The results were normally distributed and the mean values $\overline{X}$ and standard deviations σ were calculated for each group of results and were as follows:

New die, Mean $\overline{X}$ = 294°C, σ = 17°C
Old die, Mean $\overline{X}$ = 312°C, σ = 12°C

Student's T Test was used to compare these results as follows:

1. $$t = \frac{\text{Error in mean}}{\text{Standard error}}$$

2. Error in mean = Mean of old die − Mean of new die
    = 312   − 294
    = 18.

3. Standard error = $\sqrt{\text{Variance.}}$

4. Variance = $\dfrac{\text{Standard deviation}^2}{\substack{\text{Sample size for old} \\ \text{die}}} + \dfrac{\text{Standard deviation}^2}{\substack{\text{Sample size for new} \\ \text{die}}}$

$$= \frac{12^2}{50} + \frac{17^2}{25}$$
$$= 2.88 + 11.56$$
$$= 14.44.$$

5. Standard error $= \sqrt{14.44}$
   $= 3.8.$

6. $t = \dfrac{18}{3.8}$

   $= 4.7.$

7. From statistical tables for 73 degrees of freedom (i.e. sum of sample sizes $-$ 2), and $t = 4.7$, the differences are seen to be significant at the 0.1% level and recommendations were made regarding die cooling.

It is, however, useful to study the figures in more detail, since the exercise is a good example of the use of the 6σ tolerance in practice.

If we consider the old die $\bar{X} = 312°C$ and $\sigma = 12°C$

Therefore $6\sigma = 276°C - 348°C$.

The new die $\bar{X} = 294°C$ and $\sigma = 17°C$

Therefore $6\sigma = 243°C - 345°C$.

The normal distribution curves corresponding to these values are plotted in Fig. 6.7, where it will be seen that there is considerable overlap, which is shown shaded, and this could account for the fact

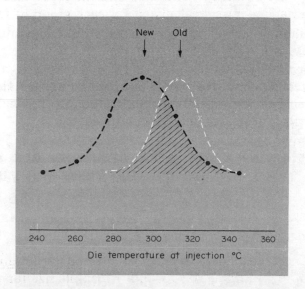

FIG. 6.7 NORMAL DISTRIBUTION CURVES FOR TWO DIE TEMPERATURES

that isolated temperature measurements appeared to show little difference between the dies. The figures also show that for both dies, the maximum temperature of castings at ejection was practically the same at ~ 345/350°C. Experience based on the temperature of other castings at ejection has shown that this temperature is rather low, so both dies would appear to be overcooled.

If we know the casting temperature at injection below which "cold" defects are produced, it is possible using statistical tables to predict the reject rate for each die. The above is an idealised concept, since in practice it is never so clear-cut, but it is a useful example and illustrates the use of statistical thinking. The predicted reject rates for the two dies at various injection temperatures are shown in Table 6.9.

TABLE 6.9 REJECT RATES FOR TWO DIES

Temperatures below which rejects are produced, °C	Reject Rate, %	
	Old die	New die
260	0	2.3
270	0	8.1
280	0.4	20.6
290	3.4	40.9
300	15.9	63.7
310	43.6	82.6
320	74.7	93.7
330	93.3	98.3
340	99.0	99.7

Table 6.9 shows very well the importance of die temperature control and the influence it can have on reject rate for "cold" defects. It also shows how rapidly the reject rate can rise if the die is critical with regard to temperature. The results also illustrate that a few random measurements of die or casting temperature are not necessarily a valid method of assessing long-term behaviour.

Case History 4—What is the Efficiency of a Furnace at the Present Time and What Would it Need to be to be Certain that it has been Improved?

Background

A foundry was interested in saving fuel on its melting units and records had been kept of metal melted and fuel used in melting copper alloys in a 6-ton HF induction unit.

Answer

The record of therms used and weight of metal melted over 67 heats was examined and a histogram was plotted. The distribution was almost normal, having a mean value $\overline{X}$ = 23.7 therms/ton and σ = 3.1 therms/ton. The above values are the basis of assessment as to whether process improvements are in fact statistically significant, e.g. if we "improve" the furnace and obtain a single figure of 19 therms/ton, does this represent a "real" improvement?

To answer this question the area under the normal distribution curve corresponding to values less than 19 therms/ton must be known, and to obtain this the normal deviate (Z) is calculated.

Z = Mean value - Experimental value

$$\overline{\phantom{\text{Mean value - Experimental value}}}$$

$$\sigma$$

$$Z = \frac{23.7 - 19}{\sigma} = 1.52$$

From tables find that $Z = 1.52$ corresponds to an area of 0.0643, and hence the probability of obtaining less than 19 therms/ton with a furnace in the standard condition is $\sim 6.4\%$. It must be assumed that the so-called "improvements" have not produced a statistically significant improvement. A figure of below $\sim$ 18.5 therms/ton would probably be significant, below 16.5 therms/ton would be significant, whilst to be certain at near the 0.1% level a figure of less than $\sim$ 13 therms/ton would be needed.

However, if by continued checking of the furnace over several heats a mean value of 19 therms/ton is obtained this can represent a statistically significant improvement.

Check over several heats and obtain a mean value of 19 therms/ton with a standard deviation of 3.1 therms/ton (i.e. the same as orginally) and apply Bessel's correction to Student's T Test. Since the sample size is rather small, it can be calculated that it would be necessary to check 7 or 8 heats to be certain of improvement at the 0.1% level.

If, however, the standard deviation σ was not 3.1 therms/ton, the number of heats to be checked could change.

The above illustrates how easy it is to draw incorrect conclusions, and this is particularly easy to do with furnace measurements, simply because such a large scatter of experimental results is produced and hence a large number are needed if the correct conclusions are to be drawn. The above case history illustrates that more information is needed after a furnace has been "improved" to enable the figures before and after to be compared on a statistical basis.

Also when considering furnace efficiency measurements it is important to know whether they are or are not normally distributed, since this could effect the statistical tests used.

It is possible that furnace efficiency figures are hardly ever normally distributed about a mean value. In fact they possibly have a skew distribution. It is probably a question of how skewed the distribution is.

This is because there is an upper limit to furnace efficiency above which the furnace cannot in practice be worked, but there is no lower limit apart from zero. A distribution curve would, therefore, have a few values at high efficiencies, the majority in the intermediate range, but a substantial number at lower efficiencies. Figure 6.8 shows the type of distribution to be expected.

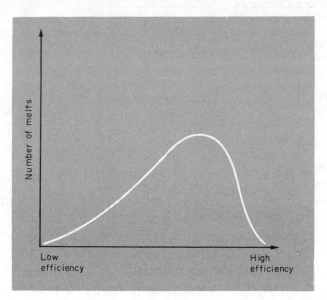

FIG. 6.8 TYPICAL DISTRIBUTION OF FURNACE EFFICIENCY

Case History 5 — Which Crucibles Shall Be Purchased?

Background

The premature failure of clay-graphite crucibles is not uncommon in foundries for various reasons. Failure can lead to metal spillage with damage to the furnace body and burner and can represent a significant down-time for the furnace.

Question

The company purchased crucibles from two manufacturers (A and B) and wished to know whether one supplier's crucibles were givng a better life than the other.

Data

In a given period 54 crucibles had been supplied by manufacturer A and 71 by manufacturer B. These had been used in three different types of furnace (X, Y and Z). The number of melts before failure for each type of furnace were compared against the crucible manufacturer and a summary of the data is give in Table. 6.10.

TABLE 6.10 NUMBER OF MELTS BEFORE FAILURE FOR DIFFERENT FURNACES AND CRUCIBLES

	Furnace X		Furnace Y		Furnace Z		Total	Total
	A	B	A	B	A	B	A	B
Melts to failure	13,15,18, 21,22,22, 24,24,25, 26,	13,26,26, 28,32,36, 38,52,58,	3,10,10, 12,13,13, 13,13,13, 18,18,20, 20,20,20, 24,24,26, 26,28,29, 30,30,30, 31,32,	8,10,11, 11,13,13, 19,21,22, 30,33,35,	13,14,14, 15,16,16, 16,17,17, 17,18,18, 18,18,18, 20,21,21, 23,23,23, 24,24,25, 25,25,26, 27,27,27, 28,29,30, 36,38,	2,6,7,8, 9,10,10, 11,13,13, 16,16,17, 18,18,18, 18,19,19, 19,20,21, 21,22,23, 24,24,25, 25,26,27, 27,29,		
Total	10	9	26	12	35	33	71	54
Mean $\bar{x}$	21.00	34.33	20.23	18.83	21.91	17.61	21.17	20.67
standard deviation σ	4.35	13.82	8.05	9.48	6.08	6.88	6.64	10.69

If we compare the above results using Student's T test we have the following situation.

1. Overall relationship manufacturers A and B

$$t = \frac{\text{Error in mean}}{\text{Standard error}}$$

$$t = \frac{21.17 - 20.67}{1.655}$$

$$t = 0.30$$

$$f = 123$$

From tables of $t = 0.30$ and $f = 123$ there is no significant difference between the manufacturers.

2. For furnace X

$$t = \frac{\text{Error in mean}}{\text{Standard error}}$$

$$= \frac{34.33 - 21.00}{4.81}$$

$$t = 2.77$$
$$f = 17$$

From tables of $t = 2.77$ and $f = 17$ there is a significant difference at the 1% level, i.e. manufacturer B is better than A.

3. For furnace Y

$$t = \frac{\text{Error in mean}}{\text{Standard error}}$$

$$t = \frac{20.23 - 18.83}{3.16}$$

$$t = 0.44$$
$$f = 36$$

From tables of $t = 0.44$ and $f = 36$ there is no significant difference between the manufacturers.

4. For furnace Z

$$t = \frac{\text{Error in mean}}{\text{Standard error}}$$

$$t = \frac{21.91 - 17.61}{1.58}$$

$$t = 2.72$$
$$f = 66$$

From tables of $t = 2.72$ and $f = 66$ there is a significant differnce at the 1% level, i.e. manufacturer A is better than B.

From the above results the conclusions to be drawn are that for furnace X, manufacturer B's crucibles should be used, for furnace Z, manufacturers A's crucibles should be used, whilst for furnace Y it is of no consequence.

No account has been taken of the purchase price, and it is possible that the crucibles having a longer life may be more expensive, and this must be taken into account.

The above case history illustrates quite well the dangers of pooling results when they represent different experimental conditions. The overall position shows no significant difference between the manufacturers, but detailed results show there are differences. In effect the significant differences for furnaces X and Z balance each other out to give an overall no significant difference.

The results can also be analysed for further information. It can be shown that for manufacturer A there are no significant differences between the crucibles used in any of the furnaces.

Similarly, it can be shown that for manufacturer B there is no significant difference between furnaces Y and Z, but that the crucibles for furnace X are significantly different.

It is possible that manufacturer B has a different "mix" for the size of crucibles used in furnace X than for furnaces Y and Z, or that there are physical differences such as crucible wall thickness which may give a stronger crucible and hence a longer life.

REFERENCES

1. S. Plummer, Precision diecasting-just how precise? *Diecasting & Metal Moulding,* June 1981
2. K. C. Bone, Quality control in diecasting, *Diecasting Engineer,* March-April 1972, pp. 34-36.
3. British Standard 6000, 1972. *Guide to the Use of Bs 6001, Sampling Procedures and Tables for Inspection by Attributes.*
4. British Standard 6001, 1972. *Specification for Sampling Procedures and Tables for Inspection by Attributes.*
5. H. L. Klein, Use of statistical tools to monitor the diecasting process, 5th National Diecasting Congress, Detroit 1968, Paper No. 143.
6. A. F. Bissell, *An Introduction to Cusum Charts.* Published by the Institute of Statisticians, London.

Index